I0001208

Martin Jacoby

Descriptions of the new Genera and Species of Phytophagous Coleoptera obtained by Dr. Modigliani in Sumatra

Martin Jacoby

Descriptions of the new Genera and Species of Phytophagous Coleoptera obtained by Dr. Modigliani in Sumatra

ISBN/EAN: 9783741198014

Manufactured in Europe, USA, Canada, Australia, Japa

Cover: Foto ©berggeist007 / pixelio.de

Manufactured and distributed by brebook publishing software (www.brebook.com)

Martin Jacoby

Descriptions of the new Genera and Species of Phytophagous Coleoptera obtained by Dr. Modigliani in Sumatra

DESCRIPTIONS OF THE NEW GENERA AND SPECIES
OF PHYTOPHAGOUS COLEOPTERA
OBTAINED BY DR. MODIGLIANI IN SUMATRA
BY MARTIN JACOBY

Since the descriptions of the *Coleoptera* of " Midden Sumatra "
published by the Leyden Museum in 1886 no similar work on
the fauna of that great island has to my knowledge appeared;
it was the more interesting to receive from Dr. Gestro the large
collection of *Phytophaga* obtained by Dr. Modigliani in Sumatra
for my examination. Nothing is of more use to the study of
Entomology than faunistic work, all isolated descriptions being
rather a hindrance then otherwise, since they have to be looked
for amongst unknown numbers of publications. Sumatra no
doubt, contains, like many other islands of the Malayan region,
numerous interesting and unknown forms, and many years must
elapse before we can hope to be acquainted in a thorough way
with its insect fauna. It may therefore be too early yet to ge-
neralize with any degree of certainty as to the affinities of the
different islands with each other and the southern parts of India,
although Wallace has attempted this and come to certain con-
clusions. As far as the *Phytophaga* are concerned, there certainly
seems ground for the assumption that Sumatra has many affinities with Burmah and other parts of India, even more perhaps
than with Borneo or Java; in regard to the genus *Sagra* however, the latter named islands seem to produce, or are the
principal home of these handsome insects, which in Sumatra are
only represented by one or two species, while India again is
rich in *Sagras*, and since these insects are of large size and
handsome appearance it is not likely that they should have been

missed by collectors in Sumatra. On the other hand, the *Eumolpidae*, *Halticidae* and *Galerucidae* are numerously represented in that island and contain striking and interesting forms, often expressed in the curiously excavated head of the male or the structure of the antennae which have frequently single enormously developed joints, or are furnished with long spikes in several species of *Galerucidae*. In the present collection the genus *Nodostoma* is largely represented, more in regard to the number of individuals than species, this and the closely allied genus *Rhyparida* are exclusively confined to the Eastern region. More than 140 new species of *Phytophaga* are here described as well as several new genera, this is more than double the number described by me in the former dutch publication, above referred to.

CRIOCERINAE.

1. **Lema capitata**, n. sp. — Black, the basal joint of the antennae, the head, thorax, elytra and the legs partly fulvous, thorax with two transverse sulci, elytra deeply and closely punctate-striate, without basal depression, the ninth row entire.

Length 2-2 $1/4$ lines.

Head finely punctured and pubescent at the vertex, the latter often with an obscure blackish spot, not raised or tuberculate but with a short and narrow groove more or less distinct, eyes large, moderately deeply notched, the clypeus and the labrum black, finely pubescent, antennae black, the basal (sometimes the two lower and the apical joint) joint fulvous, the following two joints fulvous at their base, of equal length; thorax slightly broader than long, the sides but very moderately constricted at the middle, the surface with a transverse groove at each side, nearly extending to the middle and the usual rather deep basal sulcus, the disc with a few rows of fine punctures, scutellum truncate at its apex, elytra without basal depression, deeply and closely punctate-striate, testaceous or pale fulvous; underside black, clothed with yellowish pubescence, legs fulvous, the tibiae towards the apex and the tarsi more or less black.

Hab. Balighe, October 1890. March 1891.

There are six specimens of this species before me all agreeing in the above particulars; the doubly sulcate thorax places this species near *L. lacertosa*, Lac. but the punctures and pubescence at the vertex, the obscure black spot on the latter place and the system of coloration differs quite from any Eastern *Lema* with which I am acquainted.

2. **Lema occulta**, n. sp. — Black, the head, thorax, the anterior femora and the abdomen fulvous, thorax impunctate, elytra with basal depression, fulvous, a transverse band at the base and a spot near the apex, black, the punctures strong anteriorly, more finely so posteriorly.

Var. Elytra entirely black.

Length 2 lines.

Head impunctate, fulvous, intraocular space swollen, with a central groove, the space surrounding the eyes, clothed with yellowish hairs, clypeus and labrum black, antennae long, black, the basal two joints fulvous, third and fourth joint equal; thorax slightly broader than long, the sides moderately constricted, the basal sulcus deep, the surface with a double row of very minute punctures at the middle, only visible under a strong lens, scutellum fulvous, elytra with a depression below the base, black, rather strongly punctate-striate anteriorly, and at the sides, the interstices at the latter place near the apex, costate, here and there with a few very minute punctures, the ninth row entire; the breast and the legs black, clothed with whitish hairs, the anterior femora and sometimes part of the intermediate ones, fulvous as well as the abdomen.

Hab. Benculen, April, Padang.

The typical banded form (what I take it for) and the variety look at first sight as specifically distinct but there can be no doubt that both represent the same species: in the type, the elytra are fulvous, with a black basal band which extends to about a fourth part of their length and extends across the suture to the lateral margins, a large black spot is also placed near the apex without quite extending to the sides, this form resembles much *L. Gestroi*, Jac. from Java, but in that species the antennae are entirely fulvous, the posterior spot is much larger and the abdomen is black, the thoracic sulcus is also much less deep and placed higher; the black variety of the present species agrees in every thing except the colour of the elytra, of both forms several specimens are before me.

MEGALOPODINAE.

3. **Pedrillia sumatrana**, n. sp. — Fulvous, pubescent, antennae (the basal two joints excepted) and the four anterior tibiae and tarsi black, thorax finely, elytra strongly and closely punctured. Length 1 $^1/_2$ line.

Head impunctate, fulvous as well as the labrum and palpi, eyes rather large, widely separated, antennae scarcely extending beyond the base of the thorax, black, the basal two joints fulvous, first joint rather long, curved, second short, third and fourth nearly equal, terminal joints widened, as broad as long. thorax strongly narrowed in front and at the base, the sides strongly produced into a broad oblique projection, the disc finely punctured, each puncture provided with a fulvous hair, elytra strongly but not very closely punctured, clothed like the thorax with fulvous hairs, legs robust, the femora fulvous as well as the posterior tibiae and tarsi, the four anterior ones black.

Hab. Pangherang-Pisang. A single specimen.

This is the first species of the present genus known from the Malayan region, it differs in the colour of the antennae and legs from its Eastern allies from India and Japan.

CLYTRINAE.

4. **Aspidolopha sumatrana**, n. sp. — Flavous, the antennae (the basal joints excepted) and the apex of the anterior tibiae more or less black, head dark bluish, thorax finely and closely punctured at the sides, elytra very closely and distinctly punctured.

Mas. Thorax with a central bluish spot, elytra entirely metallic blue.

Fem. Thorax unspotted, elytra flavous, a narrow transverse band at the base and a broad band below the middle, metallic blue.

Length $2\frac{3}{4}$-3 lines.

Mas. Head dark blue, nearly entirely rugose and very finely pubescent, the clypeus depressed, labrum and palpi fulvous, antennae extending to the base of the thorax, the lower four joints fulvous, the rest black, strongly widened and transverse; thorax more than twice as broad as long, the sides obliquely narrowed in front, the posterior angles rounded, the median lobe broad, but slightly produced, the surface closely punctured at the sides, the punctures of unequal size, the middle very sparingly punctured, with a V shaped dark-blue mark, the sides fulvous; scutellum thickened at the apex, the latter truncate, fulvous, the base, bluish, finely pubescent; elytra closely and rather evenly and distinctly punctured, the interstices at the sides somewhat wrinkled, underside and legs fulvous, clothed with yellow short pubescence, the lower portion of the anterior tibiae, dark blue, anterior tarsi black.

Hab. Pangherang-Pisang Octob. 1890. March 1891.

The female is larger, more convex and resembles in its pattern *A. imperialis*, Baly, but that species is larger and has an entirely impunctate thorax without blue marking.

CRYPTOCEPHALINAE.

5. **Bucharis minutus**, n. sp. — Black, the basal joints of the antennae and the legs fulvous, thorax distinctly punctured near the base only, elytra very finely punctate-striate, the base with stronger punctures.

Length $\frac{1}{4}$-$\frac{1}{2}$ line.

Of posteriorly slightly narrowed shape, shining black, the head with a few minute punctures at the vertex, the eyes large, not closely approached, narrowly and angularly emarginate at the

middle, antennae very short, only extending to the middle of
the thorax, the lower four or five joints flavous, the rest darker,
the second and third joints small, the terminal ones forming a
transversely-shaped club, thorax more than twice as broad as
long, scarcely widened, the sides evenly rounded, the angles
distinct but not produced, the surface with a few very fine
punctures, irregularly distributed, the base a little more closely
and strongly punctured, with a row of closely placed strong
punctures along the basal margin which is produced into a point
in front of the scutellum, a few similar punctures also accom-
pany the lateral margins, scutellum narrowly lanceolate, elytra
subcylindrical, not depressed below the base, finely punctate-
striate; the commencement of each row at the base marked by
deeper punctures, the interstices impunctate, the last two at the
sides slightly convex, elytral epipleurae below the shoulders but
slightly produced and rounded, prosternum subquadrate, broader
than long, its base straight, the surface strongly but sparingly
punctured, mesosternum narrowly transverse, distinctly punctured, legs fulvous or piceous with the tibiae and tarsi fulvous.

Hab. Padang, Benculen, Cauer.

One of the smallest species of this genus, of which it seems
to possess all the principal characters, the pointed median lobe
of the thorax which closely fits the narrow, not elevated scutellum and the shape of the prosternum ; the eyes are not deeply
notched and the antennae have the club six-jointed, the pygidium
is not covered by the elytra and the last abdominal segment of
the female has a shallow fovea.

6. **Bucharis oculatus**, n. sp. — Black, shining, the basal joints
of the antennae and the legs fulvous, thorax distinctly punctured
at the base only, elytra strongly and regularly punctate-striate,
the interstices at the sides convex.

Length $^3/_4$ line.

Of nearly cylindrical, posteriorly not widened shape, the eyes
extremely large, occupying the entire sides and vertex of the
head where they touch, they are but slightly and angularly
emarginate at their inner margin, clypeus flattened, rather

strongly punctured, labrum and palpi flavous, antennae extending to the base of the elytra, the lower five or six joints flavous, the others black, third, fourth and fifth joints slender, equal, the following ones slightly thickened, but longer than broad, thorax more than twice as broad as long, the sides slightly rounded and narrowed in front, the anterior portion rather strongly deflexed, the posterior margin produced at the middle into a short point, the surface very sparingly impressed with round punctures which are more distinct and closely placed at the base, forming also a transverse row near the posterior margin, scutellum narrowly lanceolate, elytra with rather strong and widely placed punctured striae which remain distinct to the apex, each row being marked at the base by a deeper impressed puncture, the interstices smooth, the last two at the sides, convex, pygidium not covered by the elytra, underside black, the prosternum broader than long, its base straight, its surface finely punctured, legs fulvous.

Hab. Pangherang-Pisang.

This equally small species may be at once distinguished from the preceding one, which it closely resembles by the very large eyes which meet at the vertex and are almost contiguous, the antennae are also longer and less thickened at the terminal joints and the elytra have deeper striae; the species cannot be the male of the preceding one, as I have both sexes of that insect before me.

7. **Bucharis constricticollis**, n. sp. — Below black, the lower part of the face, the thorax, legs and the last abdominal segment fulvous, thorax with transverse sulcus near the middle, elytra black, deeply punctate-striate.

Length $^3/_4$ line.

Of somewhat broad, subquadrate shape, the eyes occupying the entire upper portion of the head and sides, contiguous at the vertex, clypeus flavous, distinctly and rather closely punctured, antennae extending to the middle of the elytra, black, the lower three joints more or less distinctly fulvous, the third, fourth and fifth joints equal, short, the following joints more elongate but slightly widened, distinctly longer than broad, thorax twice as broad as long, the sides strongly deflexed, the lateral margins straight, not visible when the insect is viewed from above, the disc with a rather deep transverse constriction or sulcus, more strongly marked at the sides, with a few punctures near the base, the posterior margin scarcely produced at the middle, slightly pointed, accompanied by a row of deeper punctures. scutellum narrowly lanceolate, black, elytra deeply and regularly punctate-striate, the commencement of each row marked by a deeper puncture, the interstices slightly raised, especially at the sides; prosternum and legs as well as the last abdominal segment, fulvous.

Hab. Pangherang-Pisang. A single specimen.

This little species is well distinguished by the shape of its thorax, which looks much narrower, when viewed from above, the transverse sulcation placed almost at the middle and the comparatively long antennae will further assist in the recognition of this species.

8. **Melixanthus aterrimus**, n. sp. — Deep black, the basal joints of the antennae, the clypeus and the legs flavous, thorax impunctate on the disc, elytra deeply punctate-striate, the interstices convex at the sides.

Length $\frac{1}{12}$ line.

Of subquadrate, posteriorly slightly narrowed shape, the head impunctate, the eyes very large, occupying the entire sides and touching at the top of the head, clypeus deflexed, flavous as well as the labrum, scarcely perceptibly punctured, antennae extending slightly beyond the base of the thorax. the lower six joints flavous, of nearly equal length, the others black. transversely widened, thorax strongly transverse, greatly narrowed in front, the sides strongly deflexed, the anterior margin accompanied by a deep but narrow sulcus, the base near the margin with a row of deep punctures, the rest of the surface entirely impunctate, shining, the basal margin produced at the middle into an obtuse point, scutellum very elongate and narrow, elytra with deep striae which are finely punctured except at the base,

the interstices convex, especially so at the sides, pygidium finely
punctured and pubescent; below black, impunctate, the legs
flavous, prosternum subquadrate, one half broader than long,
with a few punctures, the sides with a narrow groove, the
posterior margin straight.

Hab. Pangherang-Pisang, Siboga.

This very small species may be distinguished from *M. minutus,*
Jac. (Novitat. Zoolog., 1894) by the impunctate thorax and the
black colour of its upper and under side.

9. **Cryptocephalus singularis**, n. sp. — Flavous, the antennae
black, thorax impunctate, reddish-fulvous, elytra very finely
punctate-striate, flavous, two transverse spots at the base and
a small one below the middle, black.

Length 1 ³/₄ line.

Head fulvous or (in one specimen) piceous, flat, with a few
fine punctures, eyes large, angularly notched, clypeus strongly
transverse, separated by a groove from the face, sparingly punctured, labrum and palpi flavous, antennae scarcely extending to
the middle of the elytra, black, the third joint scarcely longer
than the second, terminal joints thickened, thorax more than
twice as broad as long, much narrowed in front and widened
at the middle, the sides but slightly rounded, the surface entirely
impunctate, reddish-fulvous, the posterior margin very narrowly
black, finely serrulate, scutellum broad, flavous, narrowly margined with piceous, elytra very finely punctate-striate, the punctures widely placed but distinct to the apex, flavous, a transverse
narrow band at the base, surrounding the scutellum at which
place the elytra are raised, and a spot on the shoulders, as well
as the lateral margin below the base and the epipleurae as far
as the basal lobe, black, another small spot of irregular shape
is also placed near the apex, half way between the sutural and
lateral margin, the suture is likewise very narrowly black,
pygidium nearly impunctate, flavous, like the rest of the underside, legs pale fulvous, prosternum subquadrate, rather longer
than broad, its basal margin straight, the surface slightly depressed, finely transversely rugose.

Hab. Si-Rambé, Pangherang-Pisang.

This is a pretty little species, well distinguished by the fulvous nearly rufous thorax, the flavous elytra and their pattern.

10. **Cryptocephalus Gemmingeri,** n. sp. — Black, head distinctly, thorax extremely finely punctured, elytra strongly punctate-striate, the interstices with a row of finer punctures, black, a transverse broad band near the base, bright fulvous, underside and legs black.

Length 2 lines.

Head rather strongly but not closely punctured, black, eyes deeply notched, clypeus transversely depressed, with a few fine punctures, antennae with the basal joint piceous, the following seven joints black, the rest wanting, thorax very convex, the sides strongly deflexed, the lateral margins rather rounded, strongly narrowed in front, the posterior angles produced, the posterior margin truncate and bisinuate in front of the scutellum, the surface sparingly impressed with fine obsolete punctures, black, shining, scutellum longer than broad, its apex obtuse, the base with a small fovea, elytra narrowed posteriorly, regularly and strongly punctate-striate, the interstices with some finer punctures, arranged in somewhat irregular rows, the black colour interrupted by a bright flavous transverse band, which commencing a short distance below the base extends downwards a little below the middle, the anterior edge of this band is irregularly dentate and extends slightly upwards at the suture, the posterior margin is more regular and slightly oblique, pygidium finely punctured and pubescent, underside and legs black, prosternum broad, slightly narrowed posteriorly, the posterior angles produced, the surface finely rugose; the last abdominal segment with a deep fovea.

Hab. Sumatra (my collection).

In coloration, this species resembles *C. flavicinctus,* Jac. from Birmah, also *C. posticalis,* Jac. and several other Indian forms, from the first named it is distinguished by the entirely black thorax and the want of the subapical elytral flavous spot; *C. posticalis* has a fulvous head and thoracic margin, the position of the flavous band which does not extend to the base, separates the species from the other Indian forms.

LAMPROSOMINAE.

11. **Oomorphus nigritum**, n. sp. — Black, thorax extremely finely punctured, elytra more strongly punctured, in closely approached longitudinal rows.

Length 1 line.

Head impunctate, with a deep short sulcus in front of the eyes, the latter large, narrowly notched, clypeus separated from the face by an obsolete transverse depression, its anterior edge concave, antennae short, black, the basal two joints fulvous, the eighth joint small, terminal three joints thickened, not longer than broad, thorax strongly transverse, widened at the middle, the sides straight, much narrowed in front, posterior margin slightly sinuate at the sides, its medial lobe scarcely produced but rather pointed, the surface very finely and not very closely punctured, the punctuation a little more distinct near the base, the anterior margin accompanied by a narrow transverse groove, scutellum very small, elytra regularly convex, the apex rounded, each elytron with fifteen or sixteen rows of distinct punctures, visible to the apex but more obsolete at that place, black with a slight bluish tint, underside black as well as the legs, prosternum narrowly elongate, constricted at the middle, impunctate.

Hab. Si-Rambé, Mt. Singalang (Beccari).

Differing from *O. japanus,* Jac. in the colour of the underside and the punctuation of the elytra; this is the only species of the genus known from the Malayan region.

12. **Lamprosoma Modiglianii**, n. sp. — Narrowed posteriorly, obscure cupreous below, reddish-cupreous above, thorax finely but not very closely punctured, elytra closely and somewhat irregularly punctate-striate, legs nearly black.

Length 1 $^{1}/_{4}$ line.

Head impunctate, flat, the middle with a triangular depression, thorax strongly transverse, the sides straight and greatly nar-

rowed anteriorly, the middle widened, the median basal lobe
rather acute in front of the scutellum, basal margin sinuate at
each side, the surface finely and evenly but not very closely
punctured, scutellum small, blackish, elytra strongly convex,
narrowed posteriorly, the shoulders scarcely prominent, the basal
lobe below the shoulders angulate but not much produced, the
surface more strongly punctured than the thorax, the punctures
placed in closely approached not very regular rows which become
finer and more indistinct towards the apex, underside obscure
greenish cupreous, the legs more or less black, prosternum
elongate, narrowed at the middle, with a few fine punctures,
abdomen sparingly punctured, claws appendiculate.

Hab. Pangherang-Pisang.

This is the first species of *Lamprosoma* known from the Malayan region, China and Japan having furnished some few other
representatives of the genus. I am unfortunately not able to say
anything about the antennae of the present species, as all my
endeavours to dislodge them from their deeply placed groove
have failed and there are only two specimens available for examination; the Sumatran species resembles greatly in colour and
shape many of its South American congeners but the punctuation
of the elytra is closer and less regular than is generally the
case.

EUMOLPIDAE.

Arnobiopsis, gen n

Body elongate, subcylindrical, antennae filiform, the second
joint elongate, as long as the third, thorax subcylindrical, broader
than long, the angles dentiform, elytra broader than the thorax,
punctate-striate, femora incrassate, with a small tooth, intermediate tibiae emarginate at the apex, posterior ones entire, claws
acutely appendiculate, the anterior margin of the thoracic episternum concave; prosternum subquadrate, the base truncate.

The species for which the present genus is proposed has
somewhat the appearance and shape of a *Typophorus* or *Syagrus*,

it will however enter the first group of Chapuis' arrangement, the *Iphimeinae* on account of the structure of the thoracic episternum; *Chrysolampra* seems to be the nearest allied genus, from which and any other of that group, the present insect differs in the long second joint of the antennae and the tooth at all the femora. The structure of the antennae and the entire posterior tibiae separates the genus from *Nodostoma*.

13. **Arnobiopsis bicolor**, n. sp. — Black, the head, antennae, thorax, scutellum and the base of the femora rufous, head and thorax remotely punctured, elytra punctate-striate, the punctures very fine below the middle, the knees, tibiae and tarsi, black. Length 3 lines.

Head remotely but rather strongly punctured, the clypeus not separated from the face, its anterior edge semicircular, eyes entire, labrum fulvous, mandibles black, antennae not extending to the middle of the elytra, pale fulvous, the basal joint dilated, short, the second rather longer than the third joint, fourth and following joints elongate, not thickened; thorax one half broader than long, subcylindrical or somewhat subquadrate, the sides slightly and evenly rounded, the angles produced into a small tooth, the surface rather convex, strongly but remotely and unevenly punctured, rufous; scutellum slightly longer than broad, rufous; elytra broader at the base than the thorax, with a transverse depression below the base, strongly punctate-striate anteriorly and at the sides, the punctures very fine towards the apex, the colour deep black, shining, below and the knees, tibiae and tarsi black, the two basal thirds of the femora rufous, the anterior tibiae slightly curved and dilated at the apex, third tarsal joint deeply bilobed; prosternum subquadrate, slightly widened at the base, the latter truncate.

Hab. Si-Rambé, March and December.

The three specimens contained in this collection do not differ in any way from each other.

14. **Nodostoma atripes**. n. sp. — Black, the basal three joints of the antennae and the head more or less fulvous, thorax deeply and remotely punctured, angulate below the middle, elytra dark

fulvous, with deep basal depression, strongly punctured anteriorly, finely posteriorly, the sides at the shoulder with a short
oblique costa.

Var. Thorax and the base of the anterior femora fulvous.

Length 1 ¹/₂ line.

Head black at the vertex, fulvous at the lower portion, strongly
but remotely punctured, the epistome not separated from the
face and not more closely punctured than the head, palpi fulvous, antennae not extending to the middle of the elytra, black,
the lower three joints fulvous, third and fourth joints equal,
slightly shorter than the fifth, terminal joints slightly thickened;
thorax twice as broad as long, the sides angulate immediately
below the middle, narrowly transversely sulcate close to the
anterior margin, the surface very deeply and remotely punctured,
the punctures round and large, not more closely placed at the
sides than at the disc, the latter shining black; scutellum longer
than broad, fulvous, its apex truncate, the surface impunctate;
elytra dark reddish-fulvous, deeply transversely depressed below
the base, the latter convex, the shoulders acute, prominent and
followed by a short oblique costa, the punctures strong within
the depression and at the extreme basal margin, rest of the disc
finely punctate, below and the legs black.

Hab. Pangherang-Pisang, March and October. Si-Rambé, March
and December.

Of this small species I have five specimens for comparison,
the black head, thorax and legs in the type and the deep and
remote punctuation of the thorax, separate it from most of its
allies, the small size and other details from *N. nigripes*, Baly;
although in the variety the base of the femora and the head
and thorax are fulvous, the tibiae and rest of the legs remain
black as well as the antennae, excepting their basal three joints,
the colour of the elytra is also of a more rich fulvous than is
generally the case in many other species of this genus.

15. **Nodostoma lateripunctatum,** n. sp. — Pale fulvous, the head
impunctate, thorax angulate near the base, strongly punctured
at the same place, elytra with deep basal depression, strongly

punctured within the latter, finely towards the apex, the sides with a short oblique costa at the shoulders.

Length 1 ¹/₂ line.

Head impunctate, clypeus separated at the sides by a few punctures, antennae extending to about the middle of the elytra, flavous, the third and fourth joints equal, slender, thorax at least twice as broad as long, the sides strongly obliquely narrowed anteriorly, distinctly angulate near the base, the surface with a transverse groove near the anterior margin, sparingly but strongly punctured at the sides near the base, the rest of the disc nearly impunctate or with a few fine punctures, the posterior angles with a single long hair, elytra convex, the shoulders very prominent, with a short costa running to the lateral margin, the base with a deep transverse depression, strongly punctured within the latter, rest of the surface finely punctate-striate; femora unarmed.

Hab. Pangherang-Pisang.

This *Nodostoma* is closely allied to *N. nigritarsis*, Jac. likewise from Sumatra, but differs in the flavous colour and legs which is the same in the eight specimens before me; the strong punctures near the base and posterior angles of the thorax in connection with the other particulars pointed out above, are the principal distinguishing features of this species; the penis is short and strongly curved, the apex is not narrowed but moderately pointed, the excavated portion does not extend far upwards.

16. **Nodostoma nigrosuturatum**, n. sp. — Black, head, antennae and legs fulvous, thorax strongly punctured, angulate at the middle, the disc more or less black, the margins fulvous, elytra with basal depression, finely punctate-striate, fulvous, the margins narrowly and a broad sutural band, narrowed and abbreviated posteriorly, black.

Var. a. Elytra without black margins, the sutural band narrow or only indicated.

Var. b. Thorax fulvous, the other parts as in the type.

Length 1 ¹/₂ lines.

Head strongly but sparingly punctured at the sides, the clypeus not separated, more closely punctured, its anterior margin piceous, labrum fulvous, mandibles black, antennae long and slender, the lower four joints fulvous, the others nearly black, terminal joints slender and elongate, not thickened; thorax twice as broad as long, the sides angulate below the middle, the disc strongly but remotely punctured, with an anterior transverse groove, black, the margins more or less fulvous, scutellum fulvous, its apex broadly rounded, elytra with a distinct depression below the base, the latter convex, punctured like the rest of the surface, the punctures as usual stronger within the depression than at the rest of the disc, often surrounded by a piceous ring, the lateral margins more or less black, the suture with another broad band of variable width, gradually narrowed towards the apex; underside black, legs fulvous.

Hab. Pangherang-Pisang, Pea-Ragia, Lumut, Balighe.

This is one of the few species, in which the underside is black and therefore comparatively easy of recognition; in the seven specimens before me there is no variation in this respect but the amount of black on the upper surface is variable; *N. impressipennis,* Jac. is a closely allied species but has no black but a fulvous underside.

17. **Nodostoma nigroviride**, n. sp. — Broadly subquadrate, metallic dark green, the labrum and the base of the antennae, fulvous, thorax with strongly rounded sides, sparingly punctured, elytra with basal depression, distinctly punctured anteriorly, the sides and apex nearly impunctate.

Length 2 ½ lines.

Head with a few fine punctures, a little more strongly and closely punctured at the base of the clypeus, labrum and palpi fulvous, antennae slender, extending to the middle of the elytra, black, the lower three joints fulvous, the basal joint stained with metallic green above, terminal joints scarcely thickened, thorax rather more than twice as broad as long, of equal width, the sides strongly rounded, the angles produced, the surface with a distinct anterior transverse sulcus, rather strongly but

sparingly punctured at the disc, nearly impunctate near the
margins, scutellum subquadrate, elytra broad, subquadrate, with
a distinct transverse depression below the base, the shoulders
prominent, followed by a short costa, the punctures strong within
the depression and arranged in two rows near the suture the
rest of the disc extremely finely punctate, nearly indistinct, underside and legs nearly black.

Hab. Pangherang-Pisang.

The uniform dark metallic green colour, the rounded sides of
the thorax and the sculpturing of the latter will help to distinguish this species from others of a uniform coloration, the male
is of a little more parallel shape, the thorax has the sides
slightly angulate below the middle and the surface more strongly
punctured but as there is only one specimen of that sex before
me I am unable to say whether these differences are constant
in the sexes.

18. **Nodostoma viridissimum**, n. sp. — Metallic green, basal
joints of the antennae fulvous, the others black, thorax with
rounded sides, deeply punctured, elytra strongly punctate-striate.
the base convex, transversely depressed, tibiae aureous, tarsi
black.

Length 1 ½-2 lines.

Head coarsely and closely punctured, the interstices more or
less convex and rugose, the epistome not separated from the
face, deeply semicircularly concave at the anterior margin, palpi
fulvous, antennae extending a little beyond the base of the
elytra, black, the lower three joints fulvous, the first metallic
green above; thorax twice as broad as long, the sides rounded,
narrowed in front, the anterior angles produced into a short
tooth, the surface as deeply but not as closely punctured as
the head, with a short transverse sulcus at the sides near the
anterior margin which disappears at the middle, scutellum broad,
impunctate, its apex broadly rounded; elytra wider at the base
than the thorax, deeply transversely depressed below the base,
the latter convex, deeply punctate-striate anteriorly, the punctures closely placed but getting gradually smaller towards the

apex. the interstices shining and impunctate, slightly transversely rugose below the shoulders, underside and legs metallic green like the upper surface, the tibiae more or less aureous, the tarsi black, prosternum very broad, coarsely and remotely punctured.

Hab. Sumatra (my collection).

The colour of the antennae, rounded sides of the thorax and its coarse punctuation, colour of the tibiae will distinguish this species from any other Eastern form. There are eight specimens before me.

19. **Nodostoma bipartitum**, n. sp. — Fulvous, the antennae (the basal joints excepted) the knees, tibiae and tarsi black, thorax finely and remotely punctured, with rounded sides, elytra finely punctate-striate, with deep basal depression, the basal third portion fulvous, the rest, black, femora with a minute tooth.

Length 2 lines.

Head sparingly and finely punctured, rather strongly longitudinally strigose above the eyes, the clypeus not separated from the face, with a few punctures, reddish-fulvous, antennae only extending to the base of the elytra, black, the lower three joints fulvous, basal joint much thickened, second one half the size, third and fourth equal, the following joints widened, but longer than broad, thorax twice as broad as long, transversely convex, the sides rounded, the surface with the usual transverse groove near the anterior margin, the disc sparingly, irregularly and finely punctured, scutellum longer than broad, elytra with a deep transverse depression below the base and a short but strong, anteriorly divided costa below the shoulders, the convex basal portion nearly impunctate, the rest finely punctate-striate except within the depression and in front of the humeral callus, the black portion occupying the posterior two-thirds, not quite extending to the sides nor to the apex, it is separated from the fulvous colour anteriorly by an oblique margin, the higher portion of which is at the suture.

Hab. Si-Rambé.

This *Nodostoma* can be easily known by the arrangement of the colours and that of the legs, as well as by the non-angulate sides of the thorax, the species somewhat resembles *N. rufum*, Clark but that species is double the size and the distribution of colour is different, it has also flavous antennae and tibiae. A single, apparently female specimen was obtained.

20. Nodostoma parvulum, n. sp. — Pale fulvous, convex, thorax with strongly rounded sides, the latter deeply but remotely punctured, elytra with basal depression, strongly punctate-striate, the apex nearly impunctate.

Length 1 line.

Head rather elongate, with some fine, very remotely placed punctures, clypeus not defined, its apical margin but slightly emarginate, antennae rather long and slender, fulvous, the terminal joints slightly darker, all the joints with the exception of the second one, of nearly equal length; thorax strongly transverse, the sides rather more than usually rounded, the surface very sparingly covered with deep round punctures which are especially deep and slightly more closely placed at the sides, the anterior portion has no transverse sulcus except at the sides; elytra strongly convex, with a short transverse depression below the base, the latter punctured as well as the rest of the surface, the punctures rather strong and closely placed, the apical portion nearly smooth, the sutural and lateral margins slightly darker, at the shoulder is a short oblique costa, so frequently found in this genus.

Hab. Pangherang-Pisang, Si-Rambé.

One of the smallest species of the genus and distinguished by the very strongly rounded and not angulate sides of the thorax and its deep and remote punctuation in connection with the convex elytra.

21. Nodostoma Chapuisi, n. sp. — Flavous or fulvous, the antennae (the basal four joints excepted) black, thorax convex, the sides rounded, the disc very sparingly and finely punctured; elytra convex without basal depression, very finely punctate-striate.

Length $1\,^1/_4$-$1^1/_2$ line.

Head entirely impunctate and shining, the epistome not separated from the face, its anterior margin semicircularly emarginate, palpi incrassate, pale flavous, antennae two-thirds the length of the body, black, the lower four joints flavous, the apex of the terminal joint fulvous, thorax twice as broad as long, the sides strongly rounded, the disc convex, entirely impunctate, posterior angles produced into a slight tooth, scutellum rather narrow, elytra convex, without basal depression, extremely finely punctate-striate, the punctures surrounded by a dark ring, femora unarmed, prosternum smooth.

Hab. Si-Rambé, December.

Of this species, only two specimens are before me, the convex thorax, its rounded sides and the convex and not impressed elytra distinguishes well this species from its numerous allies.

22. Nodostoma semicostatum, n. sp. — Dark fulvous, the elytra and legs paler, head and thorax deeply and strongly punctured, the sides of the latter rounded, elytra deeply punctate-striate, the base raised, the sides longitudinally costate with a black band from base to apex.

Var. The elytral bands interrupted or entirely absent.

Fem. Elytral interstices longitudinally costate throughout.

Length 1-1 $\frac{1}{2}$ line.

Head deeply but moderately closely punctured, the clypeus not separated from the face, palpi flavous, antennae two-thirds the length of the body, flavous, third and fourth joints equal, the terminal ones very slightly thickened; thorax scarcely one half broader than long, narrowed at the base and apex, the sides rounded, the angles with a small tubercle, the surface rather convex, dark fulvous like the head and sculptured in the same way, but the punctures larger and deeper, the anterior margin followed by a narrow transverse groove, scutellum impunctate; elytra much broader at the base than the thorax, the basal portion distinctly raised with a semicircular deep depression, pale fulvous **or** flavous, very deeply punctate-striate, the sides with a longitudinal costa from the shoulder to the apex and a broad black band from the base to nearly the apex, femora unarmed, prosternum with a few deep punctures.

Hab. Si-Rambé, December.

Amongst the pale coloured species of this genus, the present one may be distinguished by the comparative long thorax, its rounded sides and deep punctures, by the equally deeply punctured elytra and the lateral costa and by the darker underside, head and thorax, in the female, all the elytral interstices are costate, there is also a small tubercle to be seen immediately below the shoulders which is however also found in several other species; in the variety the elytra are without black bands.

23. **Nodostoma Modiglianii,** n. sp. — Fulvous, the apical joint of the antennae fuscous, thorax transverse, subangulate at the middle, deeply punctured, elytra flavous with the basal portion raised, deeply punctate-striate, with a short oblique costa below the shoulders, the interstices slightly convex.

Length 1 $\frac{1}{4}$-1 $\frac{1}{2}$ line.

Head distinctly but remotely punctured, the clypeus not separated from the face with a few deep punctures, antennae two-thirds the length of the body, fulvous, the apex of the apical joints and the terminal joint entirely, fuscous, third and fourth joints equal; thorax nearly twice as broad as long, the sides obtusely angulate at the middle, obliquely narrowed before and below the latter, posterior angles tuberculate, the surface deeply and rather closely punctured, the interspaces somewhat convex and scarcely broader than the punctures themselves, scutellum small; elytra with the basal portion distinctly convex, strongly punctate-striate, the interstices sparingly and finely punctured, the sides below the shoulders with a short oblique costa, legs rather long, the femora with an extremely minute tooth.

Hab. Si-Rambé, December and March; D. Tolong, November; D. Tarabugna, Pangherang-Pisang, numerous specimens.

N. impressipennis, Jac. is very closely allied but has blackish antennae with pale basal joints and is a larger species.

24. **Nodostoma rotundicolle,** n. sp. — Fulvous, thorax impunctate, with rounded sides, elytra with deep basal depression, finely punctate-striate, femora with a distinct tooth.

Length 1 $\frac{1}{4}$ line.

Head with a few strong punctures between the eyes only, rest of the surface impunctate, the anterior margin of the clypeus but slightly emarginate, antennae two-thirds the length of the body, fulvous, the terminal five or six joints black; thorax about one half broader than long, the sides evenly rounded, the surface entirely impunctate, pale flavous or fulvous, elytra with the base strongly raised and bounded by a deep transverse depression, the shoulders prominent, limited at the sides by a row of deep punctures, posterior portion of the elytra very finely punctate-striate, femora with a small but acute tooth.

Hab. Si-Rambé, December.

This species must be separated from those of its allies, having an impunctate thorax by the rounded sides of the latter, in this respect it agrees with *N. Bohemani*, Baly but in that species the thorax is sparingly and finely punctured, the insect is larger of darker colour with the margins more or less black; specimens which belong to the female sex of the present species have a short lateral humeral costa as is often found in that sex but do not otherwise differ, the tooth at all the femora will further help to recognize the species.

25. **Nodostoma nigromarginatum**, n. sp. — Pale fulvous, the apical joints of the antennae, the breast and the sutural and lateral margins of the elytra more or less black, thorax strongly punctured at the sides only, angulate at the middle, elytra finely punctate-striate.

Var. Elytra entirely fulvous.

Length 1-1 $\frac{1}{4}$ line.

Head very sparingly punctured at the sides, a little more closely at the lower portion, the clypeus not separated, antennae extending to the middle of the elytra, fulvous, the apical three joints generally black; thorax twice as broad as long, the sides dilated, strongly rounded and subangulate at the middle, the surface dark fulvous, strongly and rather closely punctured at the sides, much more finely and sparingly so at the disc; elytra with a moderately deep depression below the base, rather strongly punctured anteriorly, the punctures getting gradually finer towards the apex but distinct at that place, the shoulders with a short oblique costa, the lateral and sutural margins black, rest of the surface pale fulvous, the breast more or less black.

Hab. Si-Rambé, December and March.

Amongst the small species of this genus, the present one may be known by the transverse thorax, its strongly produced and rounded lateral margins, angulate at the middle, in connection with its strongly punctured sides, and the dark margins of the elytra.

26. **Nodostoma varipenne,** n. sp. — Piceous, head stained with fulvous, strongly punctured, thorax short and transverse, the sides strongly rounded, the disc strongly and remotely punctured, elytra with deep basal depression, finely punctate-striate, fulvous, the margins and the disc piceous, legs spotted with fulvous, posterior femora with a minute tooth.

Var. a. Thorax and elytra nearly black.

Var. b. Thorax and elytra flavous with or without black markings.

Length 2 lines.

Head coarsely but not very closely punctured, fulvous or piceous, labrum fulvous, antennae slender, the lower three or four joints flavous, the rest piceous, fourth joint very elongate, slightly longer than the third, terminal joint ovately thickened, thorax twice and a half as broad as long, the sides strongly rounded and subangulate below the middle, the surface with a distinct transverse groove near the anterior margin, piceous, with a slight metallic gloss, strongly but irregularly and sparingly punctured ; elytra with a distinct basal depression, finely punctate-striate, the punctures indistinct below the middle, the basal portion swollen, impunctate, the sides with a short oblique costa below the shoulder, the disc pale fulvous, the sutural and lateral margins and an indistinct discoidal patch at the middle, piceous, legs fulvous, the femora more or less stained with piceous above, the posterior ones with a minute tooth.

Hab. Si-Rambé.

This is evidently one of the most variable species of *Nodostoma*

scarcely two specimens being alike ; some are testaceous above
with a slight aeneous gloss, others are of the same colour but
have the margins and a transverse band below the elytral
depression piceous, this band being sometimes widened into an
elongate patch, while a third variety is almost entirely black
above, in these specimens the base of the femora are more or
less flavous ; the species seems closely allied to *N. nigro-macu-
latum*, Lefèv. also from Sumatra, but the author has described
the head as smooth and the thorax spotted with black and
punctured at the disc only which all differs from the present
insect. The thorax of *N. raripenne* is short and transverse, scarcely narrowed in front, with very rounded and widened sides
and the punctuation is strong but irregular and remote ; it is
only in having regard to this sculpturing and shape that I am
enabled to distinguish this species from many others.

27. **Nodina fulvitarsis**, n. sp. — Black below, the basal joints
of the antennae, the apex of the tibiae and the tarsi, fulvous,
above cupreous or aeneous, thorax finely and subremotely, elytra
distinctly punctate-striate, the apex nearly impunctate.

Length ¹/₂-³/₄ line.

Of convex and broadly-ovate shape, the head very remotely
and finely punctured with a distinct ocular sulcus, the anterior
edge of the clypeus but slightly emarginate, palpi fulvous, antennae extending to the base of the elytra, fulvous or with the
terminal joints darker, third and fourth joint small, equal ; thorax
more than twice as broad as long, the sides evenly but moderately rounded, the anterior portion rather strongly deflexed,
but slightly narrowed at the sides, the surface metallic aeneous,
rather remotely and very finely punctured, the punctures at the
sides more closely placed; elytra slightly widened towards the
middle, more strongly punctured than the thorax, the punctures
arranged in regular rows, but not closely placed, the sides from
the middle to the apex, impunctate; underside and the posterior
femora black, the tibiae and the tarsi, fulvous, posterior four
tibiae distinctly notched.

Hab. Si-Rambé. Pangherang-Pisang, Padang.

The penis in this species is short and robust, strongly curved, rather strongly widened at the apex, which is obliquely pointed, the upper cavity is confined to the apical portion only. This small species is closely allied to *N. parvula*, Jac. from Burmah, but differs in the close punctuation at the sides of the thorax, which part in *N. parvula* is very finely and sparingly punctured, the sides of the thorax in the latter species also are nearly straight and the legs are entirely fulvous; there is no appreciable difference between the two sexes in *N. fulvitarsis*, as far as the punctuation is concerned.

28. Nodina brevicostata, n. sp. — Underside and the femora black, above metallic aeneous or green, the antennae, tibiae and tarsi, fulvous, thorax rather finely and remotely punctured, elytra more strongly punctate-striate, the sides with three short costae (φ).

Length $^3/_4$ line.

Head with a few fine punctures, the clypeus not separated from the face, antennae extending to the base of the elytra, fulvous, the three or four apical joints, piceous, third and fourth joints very small, equal, terminal joints distinctly thickened, thorax twice as broad as long, the sides rounded, the anterior portion rather deflexed, the surface finely and not very closely punctured, the punctures stronger at the sides, basal margin broadly produced at the middle, elytra with regular rows of rather strong punctures, distinct but finer at the apex, the sides with three short costae, scarcely extending to the middle, the spaces between them longitudinally sulcate and strongly punctured, underside and the femora black, the extreme base of the tibiae of the same colour, their apex and the tarsi, fulvous.

Hab. Doloc Tolong.

Of this species, which I at first took to be the female sex of *N. fulvitarsis*, the male is unknown to me; that the species is distinct from the last named one, of which both sexes are before me, which do not differ in the sculpture of the elytra, the examination and dissection has proved to me. *N. brevicostata* is closely allied to *N. tricostata*, Jac. from Talaut but is rather larger and differs in the black femora.

29. **Nodina nigripes**, n. sp. — Black, the basal joints of the antennae fulvous, head and thorax rather strongly and closely punctured, aeneous, elytra strongly punctate-striate, the punctures distinct to the apex, the sides with three elongate costae, not extending to the apex, legs black.

Length 1 line.

Head distinctly but remotely punctured, with a sulcus above the eyes, antennae black, the lower four joints fulvous, terminal joint strongly thickened and elongate, extending to the base of the elytra; thorax of usual transverse shape, the sides evenly rounded, scarcely narrowed in front, anterior margin not produced, posterior one nearly straight with a very slightly produced medial lobe, the surface rather strongly, evenly but not very closely punctured, the basal margin with a row of more closely placed punctures; elytra less ovate than usual, more elongate, the humeral callus prominent, the punctures regular and closely placed, rather strong, finer but distinct at the apex, the sides with three closely approached longitudinal costae from the shoulder to a little distance from the apex, the interstices between the costae rather strongly punctured.

Hab. Pangherang-Pisang.

Of this species, there are apparently only female specimens before me, it differs from *N. fulvitarsis* in being of more elongate shape, in the much longer and more acutely raised costae, placed at equal distances from each other and in the entirely black legs, the thorax is also more strongly punctured; *N. tricostata*, Jac. is smaller, has fulvous antennae and legs and a differently punctured head and thorax, and the same differences together with the more closely and finely punctured thorax separates the species from *N. indica*, Jac.

30. **Nodina sumatrana**, n. sp. — Ovately rounded, black, above aeneous, basal joints of the antennae fulvous, thorax rather closely and strongly punctured, elytra strongly punctate-striate to the apex, the interstices sparingly and finely punctured.

Length 1-1 1/4 line.

Head remotely but distinctly punctured, rather deeply obliquely sulcate above the eyes, the clypeus transverse, punctured like the head, the anterior margin deeply emarginate, palpi fulvous, antennae comparatively long, extending beyond the base of the elytra, black, the lower five joints fulvous, terminal joints transversely widened; thorax nearly three times as broad as long, the sides gradually rounded towards the apex, the surface more strongly punctured than the head, the punctures closely placed at the sides, more remotely at the disc, the posterior margin with a row of closely placed fine punctures, scutellum as broad as long, elytra convex, oblong, strongly punctate-striate, the punctures finer but distinct to the apex, the interstices very sparingly and finely punctured, those at the sides near the apex, impunctate and devoid of striae; legs and underside black, the four posterior tibiae distinctly notched at the apex, claws fulvous.

Hab. Si-Rambé.

Of less rounded shape than the majority of species and smaller than *N. gigas,* Baly, of entirely aeneous colour above and differing in the black underside and legs, the strongly punctured thorax and elytra and other particulars from other Eastern species.

31. **Nodina Balyi,** n. sp. — Broadly rounded, black, above aeneous, basal joints of the antennae fulvous, head nearly impunctate, thorax remotely, evenly and strongly punctured, elytra deeply punctate-striate throughout.

Length 1 line.

Of much broader and rounder shape than the preceding species, the head nearly impunctate at the vertex, the clypeus sparingly punctured, thorax three times broader than long, the sides strongly rounded, the anterior portion greatly deflexed, the surface remotely but strongly punctured, the punctures stronger at the middle than at the sides, elytra with widely placed punctures, regularly arranged and distinct to the apex and at the sides, underside and legs black.

Hab. Pangherang-Pisang.

Of the same coloration as the preceding species but much more rounded and broad in shape and the sculpturing of the upper parts quite different. A single specimen.

32. **Aulexis sumatrana**, n. sp. — Fuscous, breast piceous, head and thorax black, closely pubescent and finely punctured, elytra fulvous, closely and strongly punctured anteriorly, more finely so posteriorly and pubescent.

Length 3 ½ lines.

Head finely pubescent, the hairs long and grey, the vertex finely, the clypeus strongly punctured, its anterior margin straight, with two pointed teeth, labrum fulvous, eyes large, widely separated, antennae fulvous, not extending to the middle of the elytra, the third joint one half longer than the second one, the fourth as long as the two preceding joints together, terminal joints slightly shorter and thicker; thorax one half broader than long, the sides with three very small teeth, the surface with a shallow transverse depression near the base, the latter rather strongly, the anterior portion very finely punctured, clothed with long grey pubescence; scutellum subquadrate, its apex broadly truncate; elytra fulvous, without basal depression, closely and strongly punctured, especially so near the base and at the sides, breast and legs rather darker, the first joint of the anterior tarsi not longer than the second one.

Hab. Pangherang-Pisang.

This is a larger sized species than *A. nigricollis,* Baly and differs from that species in the finely not coarsely punctured head and thorax, the black colour of the head and the want of an elytral depression. There is only a single, apparently female specimen before me; from *A. Wallacei,* Baly the species may be at once distinguished by the strong punctuation of the elytra.

33. **Aulexis Wallacei**, Baly. — Obtained at Si-Rambé and Pangherang-Pisang; some of the specimens are however only half the size of the others but I am not able to separate them.

34. Demotina sumatrana, n. sp. — Obscure fulvous or piceous, clothed with curved greyish scales, sides of thorax without teeth, surface closely punctured and pubescent, elytra more strongly and irregularly punctured, clothed with grey scales and curved setae, femora with a very small tooth.

Length 1 $\frac{1}{2}$-2 lines.

Head strongly and rather closely punctured, each puncture furnished with a curved greyish hair, clypeus transverse, separated from the face by a deep groove, its surface rugosely punctured, labrum fulvous, antennae extending slightly beyond the base of the thorax, fulvous, the terminal five joints piceous or black, thickened, slightly longer than broad, the basal joint thickened, the second thinner but of equal length and longer than the following two joints, thorax about twice as broad as long, the sides very slightly rounded, entire, the anterior angles produced, the surface rather convex, strongly punctured, each puncture provided with a curved grey hair, elytra wider at the base than the thorax, rather coarsely punctured, and similarly clothed with curved scales, the punctuation more regular near the suture than at the sides; underside and legs coloured as the upper surface, prosternum broader than long, subquadrate, pubescent, the femora with a minute tooth, the anterior margin of the thoracic episternum concave.

Hab. Si-Rambé, Benculen.

A species of variable colour and size, either obscure piceous, subopaque or fulvous and more shining, according to the amount of pubescence which covers its surface, this latter is sometimes rather dense and obscures the punctuation, forming in well preserved specimens a greyish band at the sides of the thorax and several small greyish spots on the elytra, which have the pubescence also arranged in indistinct rows, in other specimens, which are probably rubbed, this arrangement is not visible or scarcely so; the species is, of course, closely allied to several other Malayan forms described by Baly but differs in details from either of them; I may add further that there are scarcely any erect stiff hairs visible on the elytra between the curved scale-like hairs; of the small spots, visible in a well marked specimen, two are placed obliquely before, and one below the middle of each elytron.

Leprotoides, gen. n.

Oblong-ovate, eyes entire, antennae filiform, the second joint short and thick, the third and fourth elongate, thorax subcylindrical without lateral margin, surface rugose, opaque, elytra closely punctate-striate, clothed with extremely short scarcely perceptible pubescence, tibiae slender and elongate, not emarginate at the apex, claws bifid, the prosternum longer than broad, subquadrate, the anterior margin of the thoracic episternum concave.

I am obliged to establish this genus for the reception of a small species of black and opaque appearance which will perhaps best enter the group of *Leprotinae* and seems allied to *Leprotes* but differs from this and any other genus of that section in having a nearly glabrous upper surface, so that only with the strongest lens can a very fine pubescence be discovered, besides the small size of the species, all other genera belonging to the group are either clothed with long hairs or covered with scales, or have the femora furnished with a tooth and the posterior tibiae emarginate at the apex.

35. **Leprotoides flavipes**, n. sp. — Black, opaque, the head fulvous, the basal joints of the antennae and the legs flavous, thorax finely rugose, subquadrate, elytra convex, widened posteriorly closely punctate-striate, the interstices costate.

Length 1-1 ¼ line.

Head broad, very finely rugosely punctured, the extreme vertex black, the entire lower portion fulvous, the clypeus transverse, separated from the face, coarsely and sparingly punctured, its anterior margin semicircularly emarginate, antennae two-thirds the length of the body, slender, black, the basal four joints flavous; thorax twice as broad as long, subcylindrical or subquadrate when seen from above, slightly narrowed in front with a very obsolete shallow depression near the anterior margin, the surface extremely closely rugose-punctate, the punctures stronger than those of the head, scutellum longer than broad finely rugose ; elytra slightly widened and very convex posteriorly with about fifteen rows of closely approached punctures, the interstices finely longitudinally costate and furnished with some extremely short grey pubescence ; underside black, nearly smooth, legs entirely flavous, the first joint of the posterior tarsi as long as the following two joints together.

Hab. Pangherang-Pisang.

36. **Heteraspis speciosa**, n. sp. — Metallic green, clothed with white pubescence, apical joints of the antennae purplish, thorax finely and closely punctured, elytra without basal depression closely punctured in semiregular rows, the interspaces finely punctate.

Length 3-4 lines.

Head rather closely and strongly punctured, especially so at the sides, where the interstices are finely strigose. the middle with a small tubercle, the anterior margin of the clypeus moderately concave, labrum metallic green with two deep punctures at the anterior edge, mandibles black, antennae extending to the base of the elytra, the lower seven joints metallic green, the others purplish, the third joint one half longer than the second but shorter than the fourth, terminal joints slightly flattened, the last one, elongate, club-shaped ; thorax one half broader than long, the sides straight as well as the posterior margin, the surface closely and strongly punctured at the sides, less closely so at the disc, sparingly clothed with white hairs, scutellum subquadrate, with a few fine punctures ; elytra not or scarcely perceptibly depressed below the base, the shoulders prominent, the punctures fine and arranged in irregular and not very closely approached longitudinal rows near the suture, the sides more closely and strongly punctured, the whole surface bright metallic green, sparingly clothed with white long pubescence ; underside and legs also metallic green and pubescent.

Hab. Si-Rambé, December.

H. speciosa is very closely allied to *H. nitidissima,* Jac. both

insects are of the same bright metallic green, but the antennae
are differently coloured and structured, the thorax in *II. nitidissima* has an anterior smooth tubercle and its posterior margin
is produced at the middle, the pubescence is also mixed with
black hairs. *II. hirta,* Fab. is much more closely punctured and
generally of a pale violaceous blue colour.

37. **Pagria sumatrensis**, Lefèv. — I have not much doubt,
that the four specimens obtained at Padang, Balighe, D. Tolong
and Siboga must be referred to Lefèvre's species who described
the type from a single specimen which represented only a variety; in normally coloured specimens the head and thorax are
greenish-aeneous, the elytra are flavous with the lateral and
sutural margin as well as two small spots below the base,
greenish or piceous, in other specimens the thorax is fulvous
and the elytral darker bands and spots more or less indistinct,
on this variety Lefèvre has founded his type, as all other particulars agree with his description. *P. bipunctata,* Lefèv. from
India seems also very closely allied but is of double the size
and has differently coloured legs.

38. **Apolepis Balyi**, n. sp. — Piceous or fulvous, pubescent,
antennae with short joints, thorax distantly punctured, each
puncture with a short curved hair, elytra punctured in rows
with similar pubescence as the thorax, the interstices at the
sides slightly costate, legs unarmed, claws bifid.

Length $^{3}/_{4}$-1 line.

Head very strongly but not very closely punctured, frontal
tubercles absent, labrum fulvous, antennae not extending beyond
the base of the thorax, black, the lower two joints fulvous, second joint as long as the first, but thinner, the following joints
shorter, terminal ones gradually thickened, thorax scarcely one
half broader than long, strongly narrowed in front, the sides
slightly rounded near the base, the surface strongly punctured
in somewhat regular longitudinal rows, the punctuation more
closely placed at the sides than on the disc, each puncture furnished with a short curved hair, elytra with a shallow depression
below the base, punctured and pubescent like the thorax, the

interstices longitudinally costate at the sides, where the punctures, are coarse and somewhat confluent, underside and legs strongly punctured, sparingly clothed with short grey pubescence, prosternum broad, rugose, not separated by a sutural groove from the episternum.

Hab. Si-Rambé.

This small species seems to possess all the structural characters of *Apolepis* except that the legs are not dentate although some slight trace of a tooth is just perceptible if carefully examined; the species differs from *A. aspera,* Baly in its more shining and much less closely punctured thorax and elytra and in the different character of the pubescence which consists of single strongly curved short hairs, not adpressed scales, the punctures are also much more distantly and regularly placed.

39. **Tricliona sulcatipennis,** n. sp. — Fulvous, the head and thorax impunctate, elytra longitudinally sulcate, the sulci scarcely punctured, forming eight or nine rows, the interstices costate at the sides.

Length 1 ½ line.

Ovately subquadrate, the head impunctate, the eyes distinctly emarginate at their inner edge, clypeus separated from the face by a distinct transverse groove, broad, impunctate, labrum and palpi fulvous, mandibles piceous, antennae extending beyond the middle of the elytra, fulvous, the second joint short and thick, the third and following joints elongate, nearly equal, thorax more than twice as broad as long, the sides strongly rounded, the posterior margin produced and rounded at the middle, the surface deflexed near the anterior angles, impunctate, shining, or very finely and sparingly punctured, when examined under a strong lens, scutellum as broad as long, elytra without basal depression, impressed with about nine rows of longitudinal grooves, which are scarcely perceptibly punctured; the interstices convex at the sides and impunctate, underside and the legs coloured like the upper surface, intermediate tibiae emarginate at the apex, anterior femora with a tooth, claws bifid.

Hab. Sumatra (coll. Dr. Kraatz and my own).

From any other species of *Tricliona* with which I am acquainted, the present one may be known by the sulcate, not punctate elytra and the nearly impunctate thorax, these differences and the unspotted colour distinguishes the species also from the other Sumatran insects belonging to this genus. I received two specimens from Dr. Kraatz in Berlin.

40. Tricliona nigrofasciata, n. sp. — Testaceous, antennae (the basal joints excepted) black, thorax finely and remotely punctured, elytra with basal depression, strongly punctate-striate anteriorly only, the lateral margin and a transverse semicircular band at the base, including a small spot, black.

Var. Entirely testaceous.

Length 2 lines.

Head impunctate, very narrowly sulcate above the eyes, the latter very large, their diameter larger than the dividing space, clypeus scarcely punctured, its anterior edge nearly straight, antennae slender, black, the basal joints more or less pale, the third joint distinctly longer than the fourth, terminal joints slightly thickened, elongate; thorax twice as broad as long, the sides rather strongly rounded, slightly narrowed in front, the anterior angles dentiform, the surface deflexed anteriorly, very remotely, sparingly and 'finely punctured, basal margin nearly straight, scutellum broad, elytra convex, parallel, with a distinct transverse depression below the base, strongly punctate-striate within the depression, the rest of the disc very finely punctured, the punctures distantly placed, the lateral margin (more broadly below the shoulders) the suture to a less extent and a ring-shaped band at the base including a round spot, black; anterior femora dilated into a strong tooth.

Hab. Padang, Siboga.

From the other two known Sumatra species described by Lefèvre, the present one is at once distinguished by the sparingly and finely punctured thorax and the design of the elytral band; the variety is entirely without markings but agrees in sculpturing etc.

41. Tricliona nigro-maculata, Lefèv. — Many specimens from

Padang. This is a most variable species in regard to coloration
of which the following varieties are before me. —

a. The disc of the thorax nearly black, the sides and the anterior margin fulvous; elytra black, a narrow stripe near the
suture and the apex, flavous.

b. Thorax with two black spots, elytra with the sides anteriorly
and a short, medially narrowed stripe from the base to the
middle, black (typical form).

c. Elytra with a narrow transverse band at the base and a still
narrower one below the middle.

d. The upper surface pale fulvous without spots.

Between these varieties there are also several intermediate degrees
of coloration. *T. fasciata,* Lefèv. is probably nothing but another
variety of the above species, agreeing nearly with the var. *c.*

42. Cleorina ornata, n. sp. — Black, the basal joints of the
antennae and the legs fulvous, head and thorax fulvous, or black,
the latter strongly punctured, elytra with basal depression, finely
punctate-striate, fulvous, the middle with a transverse black
band.

Var. Above dark fulvous.

Length 1 ½ line.

Head sparingly but strongly punctured, fulvous as well as the
labrum, eyes very large, elongate, antennae slender, extending
to the middle of the elytra, black, the lower four joints fulvous
or flavous, third and fourth joints equal, terminal joints scarcely
thickened, elongate, thorax twice as broad as long, the sides
moderately and evenly rounded, slightly narrowed in front, the
surface strongly but remotely punctured, with a transverse sulcus
near the anterior margin, elytra with the base raised, bounded
below by a rather deep transverse depression, very finely punctate-striate, with a row of deeper punctures at the sides in
front of the shoulders, the last two rows equally strong, the
interspace between raised into a longitudinal costa, legs fulvous,
the four posterior tibiae emarginate at the apex, the prosternum
broad, the anterior margin of the thoracic episternum, convex.

Hab. Pangherang-Pisang.

This species differs in coloration from most of its congeners, which are nearly all of metallic colour, *C. castanea*, Lefèv. being the only exception, but the present insect seems subject to great variation in that respect, as the thorax is either black or fulvous and the elytra sometimes devoid of the black band, this latter is broad, slightly oblique and extends upwards a little way to the suture, or the latter and the lateral margins are narrowly black.

43. Cleorina nigrita, n. sp. — Black, the base of the antennae and the legs more or less fulvous, head finely, thorax strongly and remotely punctured, elytra with basal depression, finely punctate-striate.

Fem.(?). Elytra with two short costae at the sides below the shoulders.

Length 1 $\frac{1}{2}$ line.

I must separate this species from its allies on account of the sculpture of the thorax and system of coloration, which is nearly black or bluish-black, the head is very sparingly and finely punctured, the antennae are rather long and slender and have the basal four joints entirely fulvous; the thorax is very remotely but rather strongly punctured, the punctures do not however extend to the margins, there is also the usual transverse anterior and posterior groove, the elytra have a very distinct transverse depression below the base and are finely punctate-striate, the punctures being distinct to the apex, lastly the legs are more or less flavous as well as the tarsi.

The species was obtained at Pangherang-Pisang, and I also possess it from Perak, in one of these specimens from the latter locality, the head, thorax and legs are entirely fulvous, in the other (probably the female) the elytra have two short costa at the sides, but they agree in everything else with the Sumatra specimens.

44. Cleorina (*Nodostoma*) **aeneomicans**, Baly = *antica*, Lefèv. — This seems to be again a very variable species, there are specimens before me of dark blue, aeneous, metallic green and purplish colour, and the punctuation of the thorax seems equally

subject to variation, for, although always deep and confluent at
the sides, the punctures are much more closely placed in some
specimens than in others, the head is either sparingly or more
closely punctured, the first four joints of the antennae are fulvous, the basal joint is metallic above, the elytra are rather
strongly punctate-striate and the interstices at the sides are
more or less convex; in a specimen in my collection and named
by Lefèvre the abdomen is black as this author has described it,
but in other specimens it is more or less greenish-aeneous. Baly
has placed this species in *Nodostoma* where it cannot remain on
account of the structure of the lower margin of the thorax
which is convex. Specimens were obtained at Pangherang-Pisang
by Dr. Modigliani.

45. **Cleorina malayana**, n. sp. — Below black, above metallic
dark blue, basal joints of the antennae fulvous, thorax remotely
punctured at the disc, strongly at the sides, elytra deeply punctate-striate. the interstices at the sides costate.

Length 1 ¼ line.

Head remotely but rather strongly punctured, the extreme
vertex strigose punctate, palpi fulvous, antennae slender, extending nearly to the middle of the elytra, black, the lower four
joints fulvous, the basal one stained with black above, third and
fourth joints equal, terminal ones slightly incrassate, thorax at
least twice as broad as long, convex, the sides straight, the
posterior margin broadly rounded and slightly produced at the
middle, the disc very remotely punctured, the punctures at
the sides very deep and closely placed, the anterior margin accompanied by a narrow and punctured transverse sulcus, scutellum slightly broader than long, impunctate, elytra with the
shoulders rather prominent, distinctly transversely depressed
below the base, very deeply punctate-striate, the two or three
interstices at the sides, costate, legs black as well as the underside.

Hab. Pangherang-Pisang.

This *Cleorina* is closely allied to *C. Lefèrrei, Jac.* but differs
from this and the other species described by the remote punc-

tures at the disc of the thorax and its deeply and closely
punctured sides; the elytra are also more strongly punctate-
striate; several species placed by Baly in *Nodostoma* must find
their places in *Cleorina* on account of the convex anterior thoracic episternum. These species as far as I know at present are
Nodostoma (Cleorina) basale, Baly, *N. collare,* Baly, *N. purpurei-
penne,* Baly, *N. viride,* Baly.

46. **Cleorina Gestroi**, n. sp. — Broadly ovate, dark violaceous-
blue, the terminal joints of the antennae and the tarsi, black,
head and thorax closely punctured, elytra finely punctate-striate,
with basal depression, the sides with a short oblique costa.
Length 2 lines.

Head rather strongly and closely puncture l, the clypeus not
separated from the face, its anterior edge concave-emarginate,
jaws black, antennae extending a little beyond the base of the
elytra, black, the basal joint metallic blue, the following two,
dark fulvous, third joint one half longer than the second, the
fifth and following joints rather widened and flattened; thorax
twice as broad as long, widened at the middle, the sides rather
rounded, narrowed in front, the posterior margin strongly oblique
at the sides, the median lobe broadly rounded, surface strongly
and rather closely punctured, the punctures evenly distributed,
the base with a rather deep transverse groove in front of the
lobe, scutellum subpentagonal, as broad as long; elytra broader
at the base than the thorax, subquadrate, very convex, depressed
below the base, the shoulders very prominent, followed by a
short oblique costa, the surface very finely punctured in striae,
the punctures themselves rather indistinct except those in front
of the shoulders which extend to the base, the latter without
punctures at its other portions, the punctures below the middle
nearly absent, underside and legs like the upper surface, metallic violaceous blue, tarsi black, anterior margin of the thoracic
episternum convex.

Hab. Si-Rambé.

One of the largest species of the genus and of an entirely
metallic blue colour, it is closely allied to *C. nobilis,* Lefèv. in

shape and size but that species is of reddish-cupreous colour and more strongly punctured. *C. janthina*, Lefèv. seems also very closely allied, but the short diagnosis of the author does not permit to recognize the species with certainty, the elytra are described as deeply punctured which is not the case in the present insect.

47. **Cleorina Modiglianii**, n. sp. — Metallic green or cupreous, base of the antennae fulvous, terminal joints black, head and thorax subremotely punctured, elytra with strong basal depression, strongly punctate-striate, the punctures distinct to the apex.

Fem. The sides below the shoulders with an angulate short costa.

Length 1 ¹/₂-2 lines.

Head rather sparingly punctured, the punctures strong at the vertex, very fine and sparingly at the clypeus, labrum and mandibles black, basal joints of the palpi flavous, apical joint black, antennae scarcely extending to the middle of the elytra, black, the basal joint metallic green above, the following two joints fulvous, third and fourth joints equal, terminal joints slightly widened; thorax twice as broad as long, the sides slightly rounded and narrowed in front, the surface rather convex, punctured like the head, the punctures round and remotely but evenly placed, and extending to all the margins, the latter without transverse sulci, scutellum rather broad, impunctate; elytra with a rather deep transverse depression below the base, the punctures strong, especially within the depression, the convex basal portion without punctures, excepting the row which limits the humeral callus within, the latter prominent, slightly costate below in the male, but forming a distinct angular short costa in the female, the rest of the surface finely but distinctly punctured to the apex, the extreme lateral margin reflexed and accompanied by a deep row of punctures; underside and legs metallic green, tarsi purplish.

Hab. Si-Rambé, Pangherang-Pisang.

This species varies rather considerably in its coloration, which is either metallic green, or green with a dark purplish band at the sides of the elytra or entirely cupreous, but structural differences I am unable to find; the comparatively sparingly punctured head and thorax and the distinct humeral costa separates the species from its congeners.

48. **Corynodes brevipennis**, n. sp. — Short, subquadrate, metallic green, antennae purplish, thorax strongly transverse, very sparingly and remotely punctured at the disc, elytra with basal depression at the sides, finely and regularly punctate-striate, claws appendiculate.

Length 3 lines.

Head with a very deep triangular depression, within which a small tubercle is placed, the vertex remotely but distinctly punctured, clypeus separated by a deep groove from the face, transverse, very finely rugose, labrum metallic green, antennae only extending to the base of the elytra, the lower joints dark greenish, the terminal five joints purplish, broadly flattened, thorax twice as broad as long, widened at the middle, the sides rounded, scarcely narrowed in front, the anterior and posterior margin accompanied by a narrow transverse groove, the surface very sparingly and remotely but distinctly punctured at the disc only, the sides and base nearly impunctate, scutellum slightly longer than broad, its apex rather pointed, impunctate, elytra with a transverse, rather shallow depression at the sides, finely but not closely punctate-striate, the sides with a blunt tubercle below the shoulders, underside metallic green; tarsi purplish, claws appendiculate.

This species, of which I possess a single apparently female specimen only from Sumatra, may be known by its general short and subquadrate shape and the transverse and very sparingly punctured thorax as well as by the fine and regularly punctured elytra.

49. **Colaspoides nigripes**, Jac. — Notes Leyden Mus. Vol. VI, 1884; Lefèvre, Notes Leyden Mus. Vol. IX, 1887.

Specimens from Siboga, Pangherang-Pisang, Si-Rambé.

Apparently a very variable species in regard to colour, several varieties have been described by Lefèvre; in the specimens contained in the present collection, the head is rather strongly and closely punctured with a more or less deep longitudinal fovea, and the legs are either piceous or dark fulvous, as well as the underside; in the type the legs are black; some small differences are perceptible in regard to sculpture in the Sumatran specimens obtained by Dr. Modigliani but I do not think these to be of specific value but due. to local variation.

50. **Colaspoides Modiglianii**, n. sp. — Metallic cupreous, dark blue or aeneous, the antennae fulvous, the terminal joints black, thorax very remotely punctured, elytra not depressed at the sides, strongly and closely punctate-striate, interstices costate at the apex, posterior femora with a large tooth, anterior ones with a small one.

Length 2-2 $^1/_2$ lines.

Mas. Head with a few fine punctures, widely separated, clypeus broad, with rounded sides, separated above by a transverse depression, sparingly punctured, labrum fulvous, mandibles black, antennae extending beyond the middle of the elytra, fulvous, the last two and part of the third joint, black, all the joints, the second one excepted, of equal length or very nearly so, thorax twice as broad as long, the sides rounded, the anterior angles acutely produced, the surface very irregularly, but distinctly and sparingly punctured, the margins and sides impunctate, elytra without depression below the shoulders, strongly punctured in closely approached rows, distinct to the apex, the interstices at the sides below the shoulders, somewhat wrinkled, those at the apex, costate, the underside and the femora more or less metallic bluish or aeneous, tibiae and tarsi dark fulvous, the posterior tibiae with a large triangular tooth; penis short and strongly curved anteriorly, broadly rounded towards the apex, the latter armed with two teeth or points, one at each side.

Fem. Head and clypeus very minutely punctured, thorax less distinctly punctate, interstices at the sides of the elytra more or less longitudinally costate.

Hab. Padang.

Several very closely allied species of *Colaspoides* from the East
have been described, all possessing a strongly developed tooth
at the posterior femora, but the present species does not agree
with either of them. I will here point out the differences between
this and the allied species; the latter are *C. glabrata*, Jac., *C.
apicicornis*, Jac., *C. varians*, Baly, *C. laevicollis*, Lef., *C. speciosa*,
Lef. and *C. ciliatipes*, Lefèv. of these, *C. glabrata* is a much
larger insect without raised elytral interstices, *C. apicicornis* has
geminate-punctate elytra, *C. varians* is larger and has a closely
and distinctly punctured thorax, *C. laevicollis* has distinctly transversely depressed elytra, but the author says nothing about the
interstices or the sex. *C. speciosa* is described with fulvous antennae and of $8\,^{1}/_{2}$ mill. length. *C. venusta*, Lefèv. has a sub-
basal depression and sub-geminate punctate elytra, lastly *C. ciliatipes* is 3 lines in length, has aciculate thoracic punctures and
long and dense tibial pubescence at the apex; the structure of
the penis in the present species seems another mark of distinction
although nothing has been said about this organ in the allied forms.

51. **Colaspoides laevicollis**, Jac. — Lefèvre's name having been
overlooked by me at the time of publication (Genoa Annals, XXVII.
1889) I alter the name to *C. glabricollis*.

52. **Aulacia flavifrons**, n. sp. — Piceous or blackish, the head,
the basal joints of the antennae and the legs fulvous or flavous,
thorax rather closely punctured, elytra punctured in irregular
rows anteriorly, regularly posteriorly, the interstices at the sides
costate.

Length $^{3}/_{4}$-1 line.

Head with a few extremely minute punctures, the vertex with
a central longitudinal groove and deeper lateral grooves which
unite in front. clypeus wedge-shaped, distinctly punctured,
labrum and palpi fulvous, antennae extending to the middle
of the elytra, flavous, the terminal four joints black, the basal
two joints thickened, third and following joints nearly equal,
slender, apical ones thickened. thorax more than twice as
broad as long, slightly widened at the middle, the sides rounded
and strongly narrowed in front, the surface rather closely and

finely punctured, scutellum broad, impunctate, elytra more strongly punctured than the thorax, the punctures rather irregularly and widely placed at the anterior portion, but in regular rows posteriorly, the outer two interstices longitudinally convex, legs fulvous, the posterior femora sometimes piceous, the intermediate tibiae not emarginate at the apex.

Hab. Siboga and Pangherang-Pisang.

Closely allied to *A. diversa*, Baly, but the antennae with dark apical joints, the thorax closely not remotely punctured, the punctuation of the elytra not arranged in double rows and the intermediate tibiae entire.

53. **Aulacia ornata**, n. sp. — Ovately-rounded, fulvous, the head and thorax impunctate, the sides of the latter and two spots at the base, black, elytra remotely and semiregularly punctate-striate, black, the disc with an oblique large fulvous patch, intermediate tibiae entire.

Length 1 line.

Head as in the preceding species, impunctate, the clypeus broader, indistinctly separated from the face, impunctate, slightly concave, antennae extending beyond the middle of the elytra, entirely fulvous, all the joints slender with the exception of the first two, thorax shaped as in the preceding species, impunctate or with a few fine punctures at the sides only, fulvous, the sides narrowly and two obscure patches or spots at each side of the base, black, scutellum fulvous, elytra rounded, very remotely punctured in semi-regular rows anteriorly which become more regular and singly towards the apex, the black portion interrupted by a longitudinal fulvous band which extends from below the base to the suture at some distance from the apex, underside dark fulvous, legs paler.

Hab. Pangherang-Pisang.

Allied to *A. bipustulata*, Baly but with entirely fulvous antennae, the thorax differently coloured and the elytra not closely but remotely punctured, the punctures not geminate anteriorly and the intermediate tibiae entire, not emarginate; the elytral spot is also of elongate not rounded shape.

CHRYSOMELINAE.

54. Chrysomela malayana, n. sp. — Metallic greenish-aeneous, antennae bluish-black, thorax remotely and finely punctured, the sides foveolate-punctate, elytra strongly punctured in distant irregular rows, the interstices very minutely punctate.

Length 3-4 lines.

Of subquadrate, convex and parallel shape, dark greenish-aeneous, the head with a few minute punctures, the clypeus separated by a deep triangular groove, labrum blackish, palpi thickened, the terminal joint longer than the preceding one, antennae bluish-black, extending to the base of the elytra, the terminal five joints thickened, longer than broad, thorax twice as broad as long, but little widened at the middle, the sides nearly straight, slightly rounded anteriorly, the angles acute but not produced, the surface irregularly and finely but distinctly punctured at the disc, the interstices very minutely punctate, the sides not thickened but with some very deep punctures at the base and at the anterior angles, scutellum smooth, not longer than broad, elytra regularly convex, not dilated posteriorly, with distantly and irregularly placed rows of punctures, which at the sides approximate in pairs, the interstices also sparingly and finely punctured; underside and legs coloured as above, nearly impunctate.

Hab. Pangherang-Pisang. Two specimens.

Two other species of true *Chrysomela* are at present known from Sumatra, *C. stictica*, Stål and *C. sumatrensis*, Jac. The present species differs entirely from the first named in the sculpture of the elytra and from the last, in the general coloration and puncturing of the same part.

HALTICINAE.

55. Acrocrypta basalis, n. sp. — Black, the apical joint of the antennae yellowish-white, thorax very short and transverse, finely punctured, elytra more strongly and somewhat rugosely punctured, fulvous or flavous, the base with a transverse black band.

Length 2 lines.

Head with a few very fine punctures, the vertex sometimes fulvous, frontal elevations scarcely raised, broad, nearly contiguous with the broad clypeus, the latter strongly raised, with a few minute punctures, labrum flavous, palpi strongly incrassate, pale, stained with piceous, antennae extending to the base of the elytra only, black, the last joint whitish, first joint elongate, second one smaller than the third, the following joints thickened, pubescent, longer than broad, thorax three times broader than long, the sides nearly straight, with a narrow margin, the anterior angles broadly oblique, much thickened, the surface irregularly and finely punctured, black, posterior margin sinuate at the sides, nearly truncate at the middle, elytra strongly convex, rather strongly and somewhat rugosely punctured, the basal portion to about one fourth the length, black, the rest fulvous or flavous, underside and legs black, the first joint of the posterior tarsi as long as the two following joints together, claws appendiculate, anterior coxal cavities closed.

Hab. Si-Rambé, Pangherang-Pisang.

Closely allied to *A. dimidiata*, Baly but smaller, the antennae with the apical joint pale and the abdomen black, the elytra have the black band at the base extending some distance downwards along the lateral margin, a single specimen from Pangherang-Pisang is rather larger, darker in colour and the elytra are more finely and not rugosely punctured, but otherwise there is no difference, in the absence of more specimens I am unable to say whether this is a variety or another closely allied species.

56. **Podagrica rotundata**, n. sp. — Ovately-rounded, light brown, thorax extremely finely punctured with a perpendicular groove at each side, elytra regularly punctate-striate, the interstices longitudinally convex.

Length 2 lines.

Of rounded and somewhat flattened shape, entirely light brown,

the head impunctate without frontal elevations, clypeus broad,
impunctate, the space above the base of the antennae with an
oblique groove, palpi rather slender, the penultimate joint but
slightly incrassate, the terminal one acutely pointed, antennae
not quite extending to the middle of the elytra, entirely fulvous,
all the joints with the exception of the second one, of nearly
equal length, thorax more than twice as broad as long, the
sides nearly straight, the anterior angles somewhat truncately
oblique with an obsolete depression behind the eyes, the posterior
margin with a broad but slightly produced median lobe and a
distinct perpendicular groove at each side, the surface extremely
minutely and closely punctured, scutellum small, elytra wider
at the base than the thorax, rather strongly and regularly
punctate-striate, the interstices especially at the sides, longitudinally costate, their epipleurae very broad, posterior tibiae with
a distinct spine, prosternum and mesosternum rather broad.

Hab. Si-Rambé, March and December.

In its broad and rounded shape and the costate elytral interstices, this species resembles much a species of my genus *Erystus,*
but the nearly straight sides of the thorax and the distinct perpendicular groove of the latter, place the insect in *Podagrica*
and not in *Nisotra* which two genera have been more clearly
defined by Weise (Deutsche Ent. Zeitsch., 1892). Amongst the
numerous exotic forms which exist, it is however very difficult
and often impossible to define the limits of a genus to a satisfac-
tory degree while classification is comparatively easy with a
few species only available for examination and unless the mo-
nographer has a great material to work from, no certain arrangement can be arrived at.

57. **Crepidodera minuta**, n. sp. – Flavous, the apical joints of
the antennae, black, thorax deeply transversely sulcate, strongly
but sparingly punctured, elytra rather strongly punctate-striate,
the interstices flat.

Length ½ line.

Of narrow, parallel shape, the head impunctate, the frontal
elevations obsolete, the clypeus very narrowly transverse, an-

tennae scarcely extending to the middle of the elytra, flavous, the two or three terminal joints piceous or black, the last five joints distinctly thickened, not longer than broad, the last joint more elongate, pointed, second joint thickened, as long as the following two joints together, the next three joints, longer and thicker again, thorax one half broader than long, the sides strongly narrowed near the base, much rounded and widened near the apex, the anterior angles acute but not produced, the surface sparingly but strongly punctured, the basal sulcus deep and placed close to the basal margin, but not quite extending to the lateral margins, scutellum small, elytra not depressed below the base, distinctly punctured to the apex, the punctures rather larger at the base, closely placed and arranged in regular rows, underside sparingly punctured, legs short and robust, the posterior femora scarcely more thickened than the others, the posterior metatarsus, scarcely longer than the second joint, claws appendiculate, prosternum narrowly elongate, closing the anterior coxal cavities.

Hab. Padang.

This is the smallest species of the genus known to me and almost resembles a species of *Corticaria* in shape and colour, the terminal joints of the antennae are distinctly incrassate, almost moniliform in shape, but I see no reason to separate the species from *Crepidodera* for the present.

58. Manobia dimidiaticornis, n. sp. — Blackish, antennae and anterior legs flavous, 8th and 9th joint of the former, black, thorax finely and sparingly punctured, the basal groove deep, elytra strongly punctate-striate, the base swollen.

Var. The antennae and the entire upper surface pale fulvous.

Length 1 line.

Head impunctate, black, the frontal tubercles small, narrowly oblique, palpi fulvous, antennae extending to the middle of the elytra, flavous, the eighth and ninth joints black, the second joint thickened, as long as the third, the following three joints equal, slightly shorter than the seventh joint, thorax subquadrate, one half broader than long, the sides straight, the anterior angles oblique and thickened, posterior margin broadly produced at the middle, the basal sulcus rather deep, strongly punctured, the rest of the surface very sparingly and finely punctate, elytra with the basal portion strongly raised, the shoulders prominent, the surface strongly punctate-striate, the interstices costate at the sides, legs fulvous, posterior femora piceous.

Hab. Si-Rambé.

This species may be known from any of its allies by the coloration of the antennae in connection with the black shining upper surface; it is closely allied to *M. pallipes,* Jac. but in that species the antennae have four intermediate black joints and the thorax is impunctate ; I must look upon entirely pale fulvous specimens in regard to the antennae and upper surface, as varieties, since I am unable to find any differences in structures, they were also obtained at the same localities and at the same time.

59. **Lactica transversicollis**, n. sp. — Bluish-black, the head, the basal joints of the antennae and the thorax fulvous, the latter with distinct transverse sulcus, microscopically punctured, elytra dark metallic blue, extremely finely and closely punctured in semi-regular rows.

Length 1 line.

Head impunctate, the frontal tubercles nearly obsolete, carina short and blunt but distinct, palpi slender and pointed, antennae extending to the base of the elytra, black, the lower four joints fulvous, the second and the following two joints nearly equal, the other joints slightly longer, finely pubescent, thorax transverse, twice as broad as long, the sides nearly straight, the anterior angles oblique, the surface convex with a distinct slightly sinuate transverse sulcus, bounded laterally by a perpendicular groove, the disc extremely minutely punctured, only visible under a strong lens, scutellum small, black, elytra subcylindrical without basal depression, extremely closely punctured in indistinct longitudinal rows, the interstices also very minutely punctate, underside and legs blackish or the latter obscure fulvous at the base of the tibiae, the first joint of the posterior tarsi as long as the second one, prosternum longer than broad, impunctate.

Hab. Padang, Si-Rambé, Cauer.

A small species, allied to *L. sumbawaensis*, Jac. but with differently coloured underside and distinct thoracic sulcus, the punctuation of the thorax is scarcely perceptible and that of the elytra, although distinct, so close that the arrangement of the punctures in rows is only apparent here and there, when the insect is viewed in certain lights.

60. **Longitarsus Gestroi**, n. sp. — Testaceous the labrum, the scutellum and the posterior femora piceous, thorax and elytra nearly impunctate.

Length 1 ³⁄₄ line.

Head impunctate, the frontal elevations obsolete, transverse, clypeus rather broad, distinctly raised and narrowed between the antennae, labrum black, palpi slender, flavous, antennae scarcely extending to the middle of the elytra, flavous, the terminal joints more or less stained with piceous, the second joint half the length of the first but only a little shorter than the third joint, the following joints equal, scarcely thickened, thorax one half broader than long, the sides straight, the anterior and posterior angles oblique, the surface rather convex, impunctate, scutellum black or piceous, broader than long, elytra nearly parallel, not perceptibly punctured but covered with small piceous spots, wings present, the sides of the breast and the posterior femora piceous, posterior tibiae strongly widened towards the apex and broadly chanelled, with a distinct spur, the first joint of the posterior tarsi, scarcely longer than the following three joints together, the first joint of the anterior tarsi widened in the male.

Hab. Benculen, April 1891.

This *Longitarsus* resembles many European and Asiatic species in coloration, but may be known by the dark scutellum and the comparatively short metatarsus of the hind legs, which is however still longer than in most species of *Aphthona;* the last abdominal segment of the male has a central very narrow groove extending through its entire length.

61. **Longitarsus sumatrensis**, n. sp. — Below piceous, antennae

black, the basal joints fulvous, thorax subquadrate, impunctate, fulvous, elytra obscure fulvous, punctured in closely approached distinct rows, apex of the posterior femora blackish.

Length $1/2 - 3/4$ line.

Head obscure fulvous, impunctate, the vertex minutely transversely strigose, when seen under a strong lens, frontal tubercles absent, labrum and palpi piceous, antennae long and slender, two-thirds the length of the body, black, the lower three joints fulvous, the third joint slightly longer than the second, but distinctly shorter than the fourth joint, thorax one half broader than long, the sides straight, very slightly widened towards the apex, anterior angles obliquely thickened, the surface convex, obscure fulvous, the sides with a few minute punctures, only visible under a deep lens, the disc impunctate, elytra slightly widened towards the middle, of the same colour as the thorax, finely but distinctly punctured in closely approached rows, wings present, underside piceous, legs fulvous, the apex of the posterior femora, black, posterior tibiae with a row of teeth at the upper margin, their metatarsus half their length.

Hab. Padang.

A small species, which may be known by its obscure fulvous colour, the black and long antennae and the regular punctuation of the elytra.

62. **Longitarsus annulicornis**, n. sp. — Flavous, the antennae as long as the body, each joint fuscous at the apex, thorax scarcely perceptibly punctured, elytra closely and semiregularly punctured, the suture very narrowly blackish.

Length 1 line.

Of elongate, parallel shape, the head finely transversely wrinkled at the vertex, the frontal tubercles indistinct, bounded by a transverse groove behind, carina acutely raised, extending to the apex of the clypeus, labrum piceous, antennae extending to the end of the elytra, flavous, all the joints from the fourth upwards, fuscous at the greater portion of the apex, third joint half the length of the fourth, thorax one half broader than long, the sides straight, the anterior angles oblique, the surface scarcely visibly punctured, when seen under a strong lens, elytra finely and very closely punctured, the punctures somewhat arranged in regular rows here and there, the suture very narrowly black, underside and the femora slightly darker, the posterior femora long and stout, extending to the end of the abdomen, upper edge of the posterior tibia finely dentate, their metatarsus half their length.

Hab. Doloe Tolong.

A single specimen only of this interesting little *Longitarsus* was obtained, it will be known by the long antennae and their colour.

63. **Longitarsus Balyi**, n. sp. — Entirely pale fulvous, head and thorax impunctate, the latter broader than long, elytra not perceptibly punctured, with rows of small piceous dots.

Length $^1/_2$ line.

Head impunctate, the frontal tubercles narrowly oblique, distinct, carina very short and rather thick, clypeus moderately concave at its anterior edge, antennae extending slightly beyond the middle of the elytra, pale fulvous, the second joint thickened, scarcely shorter than the third one, the following joints slightly more elongate, thorax about one half broader than long, the sides rather evenly rounded, the anterior angles (to a small extent) oblique, the surface rather convex, entirely impunctate, elytra with scarcely a trace of punctures but with rows of small piceous spots, indicating the punctures, underside also impunctate, prosternum slightly longer than broad, the first joint of the posterior tarsi as long as the following joints together.

Hab. Padang.

A small species, distinguished by the nearly impunctate upper surface and the comparatively short metatarsus of the posterior legs.

64. **Aphthona nigrita?** Motsch. — Specimens from Pangherang-Pisang agree almost entirely with those from Ceylon contained in my collection and which I refer to Motschulsky's species, although this author's descriptions are practically useless. The Sumatra specimens are black, with the exception of the antennae

in which only the terminal three or four joints are of that colour, the rest are flavous as well as the anterior legs and posterior tibiae; the upper surface is deep black and shining, the head is impunctate, the eyes are very large, the frontal tubercles are small and oblique, the carina distinct, the antennae are long and slender, the thorax is subquadrate, the sides slightly rounded, the surface entirely impunctate, the elytra are scarcely perceptibly punctured.

65. **Sphaerometopa sexmaculata**, n. sp. — Fulvous, the antennae (the basal and apical joint excepted) black, thorax scarcely perceptibly punctured, elytra flavous, finely and closely punctured, a spot on the shoulder, a transverse band at the middle and another near the apex, black.

Length 2 ½ lines.

Head impunctate, the frontal tubercles small but distinct, clypeus broad, impunctate, palpi strongly incrassate, antennae extending nearly to the middle of the elytra, black, the basal three joints fulvous, the terminal one yellowish-white, third joint one half longer than the second, the following joints slightly widened, pubescent, thorax three times broader than long, short, but slightly widened at the middle, the anterior angles oblique, thickened, the surface rather convex, scarcely perceptibly punctured, fulvous like the head, scutellum triangular, fulvous, elytra very closely, finely and irregularly punctured, flavous, the shoulders with a black spot, the middle with a narrow but regular transverse black band not quite extending to the suture, another more irregular shaped band is placed at a little distance from the apex, somewhat oblique, it extends to the suture, where it is slightly widened, but not quite to the lateral margins, underside and legs fulvous, tibiae slightly darker at their outer edge, the first joint of the posterior tarsi as long as the following joints together.

Hab. Si-Rambé. A single specimen.

Distinguished from any of its allies by the pattern of the elytra.

66. **Sphaerometopa imitans**, n. sp. — Dark fulvous, the antennae (the basal and apical joint excepted) and the tibiae and

tarsi black, thorax very minutely punctured, elytra black, with a large flavous patch before, and another below the middle, finely punctured.

Length 3-3 ¹/₂ lines.

Of broadly ovate and convex shape, the head impunctate, fulvous, frontal tubercles distinct, transverse, clypeus strongly swollen, eyes very large, palpi incrassate, antennae not extending to the middle of the elytra, black, the basal joint fulvous, the apical one yellowish-white, third joint one half longer than the second, following joints more elongate and slightly thickened, thorax nearly three times broader than long, fulvous, the sides nearly straight, the anterior angles obliquely thickened, anterior margin straight, posterior one widened at the middle, surface scarcely perceptibly punctured, scutellum broad, fulvous, elytra scarcely more strongly punctured than the thorax, pale flavous, this colour divided into four large patches by a narrow transverse black band at the middle, a similar band extends across the base, the suture and lateral margins are likewise narrowly black, below dark fulvous, the tibiae and tarsi blackish, posterior femora moderately incrassate, their tibiae armed with a strong spine, the first joint of the posterior tarsi as long as the following three joints together, claws appendiculate, mesosternum elongate, widened at the base, anterior coxal cavities open.

Hab. Si-Rambé.

S. imitans can easily be mistaken for two other Sumatran species which almost entirely resemble it in coloration but belong to the genus *Imolia,* Jac. on account of the closed coxal cavities. In the present insect the latter are open and the antennae are slightly dilated but they are not typical nor is the structure of the metatarsus of the hind legs which is much longer than in *S. acroleuca,* Wied. the typical form, this however is a difference of degree only, the shape and other structural characters agree with the genus. *S. ornata,* Baly has closed cavities and has been placed by me in *Imolia,* it differs besides in the distinctly punctured thorax and elytra, the narrow flavous bands of the latter and the entirely fulvous legs.

67. **Sphaerometopa obsoleta**, n. sp. - Black, terminal joint of the antennae whitish, thorax extremely finely punctured, elytra obsoletely punctate and finely wrinkled, fulvous, the lateral margin and a small spot attached to it below the shoulders, black.

Length 2 ³/₄ lines.

Of oblong and convex shape, the head impunctate or nearly so, deeply transversely grooved between the eyes, the frontal tubercles narrowly transverse, clypeus broad, lower portion of face more or less fulvous, palpi incrassate, antennae short, black, the apical joint yellowish-white, third joint twice as long as the second one, following joints gradually widened, terminal joint more elongate, thorax very short and transverse, of usual shape, widened at the middle, the sides nearly straight, the anterior angles obliquely thickened, the surface extremely minutely punctured, black (sometimes stained with fulvous) scutellum elongate, piceous, elytra fulvous, the punctuation extremely fine and obsolete, the interstices uneven or finely wrinkled, the lateral margin narrowly black, with a (sometimes obsolete) small spot near the middle, underside and legs black.

Hab. Pangherang-Pisang.

Very closely allied to *S. 4-punctata*, Jac. from Borneo but without the spots; the lateral one excepted, and principally differing in the obsolete and extremely fine elytral punctuation which in *S. 4-punctata* is very distinct, the antennae in that species also have the lower three joints fulvous. Of the genus *Sphaerometopa*, distinguished by the short and dilated antennae, the narrow and transversely shaped thorax in connection with the open cotyloid cavities, there are now nearly a dozend species know, of two others, from the neighbouring islands of Borneo and Java contained in my collection I give here the descriptions.

68. **Sphaerometopa nigropicta**, n. sp. — Fulvous, antennae black, the apical joint yellow, head and thorax fulvous, finely and closely punctured, elytra more strongly and very closely punctured, obscure testaceous, the margins and four spots placed transversely below the middle, black.

Length 3 ½ lines.

Head with some very fine punctures, the frontal elevations oblique, joined to the broad clypeus, the latter finely punctured, antennae short, black, the lower three joints fulvous, the apical one yellowish, thorax of the same shape as in the other species of the genus, finely and closely punctured, scutellum fulvous, elytra more distinctly punctured than the thorax and as closely so, testaceous, all the margins narrowly black, each elytron also with two black spots, placed transversely, immediately below the middle, underside and legs fulvous.

Hab. Borneo (my collection).

Of nearly similar coloration as *S. 4-punctata,* Jac. but larger, more oblong in shape, the margins of the elytra black and the spots placed below, not *at* the middle as in the allied species, where they are also arranged slightly oblique instead straight as in the present insect.

69. **Sphaerometopa Fruhstorferi,** n. sp. — Fulvous, the antennae (the basal and apical joints excepted) the legs and sides of the breast, black, thorax finely and sparingly punctured, elytra more strongly but not closely punctured, bluish-black, the sutural and lateral margins narrowly fulvous.

Length 3 ½-4 lines.

Head fulvous, nearly impunctate, frontal tubercles and clypeus as in the other species of this genus, mandibles fulvous, the apex black, antennae more elongate than usual, black, the lower four joints fulvous, the apical two joints yellowish, third joint one half longer than the second, fourth, longer than the third joint, intermediate joints but slightly thickened, thorax three times broader than long, not much widened at the middle, the sides rounded near the middle, the anterior angles obliquely thickened, the surface rather convex, finely and sparingly punctured at the sides, the disc nearly impunctate, scutellum fulvous, elytra much more strongly punctured than the thorax, the punctures not very closely placed, bluish-black, the margins and their epipleurae fulvous.

Hab. Java (Fruhstorfer).

At once to be distinguished by the bluish colour of the elytra, rather long antennae and large size. I received three specimens of this fine species from Mr. Fruhstorfer.

70. **Sphaerometopa pallidipennis**, n. sp. — Black, the apical two joints of the antennae pale yellow, thorax finely and sparingly punctured, elytra pale flavous, the extreme lateral margin piceous, the disc closely and distinctly punctured.

Length 3 lines.

Head impunctate, black, frontal elevations narrowly oblique, joined to the broad clypeus, labrum large and prominent, palpi incrassate, black, extending a little beyond the base of the elytra, the apical two joints yellowish white, the third joint, one half longer than the second, but slightly shorter than the fourth joint, terminal joint, shortened and moderately widened, thorax three times broader than long, the sides nearly straight, the anterior angles obliquely thickened, the surface very finely and sparingly punctured, black, scutellum black, elytra convex, more strongly and closely punctured than the thorax, with a small depression at the sides below the shoulders, pale flavous, the extreme lateral margin and the epipleurae more or less stained with piceous, underside and legs black.

Hab. Si-Rambé.

Smaller than *S. nigricollis,* Duviv. and with black scutellum; the elytra pale flavous instead of reddish-brown and the thorax more finely punctured, the sides of the elytra as well as the apex also obsoletely piceous or pale brownish. *S. intermedia,* Jac. has entirely black antennae with shorter and wider joints and a fulvous scutellum, *S. obsoleta,* Jac. differs in the nearly impunctate thorax, in the antennae having the last joint pale only, and in the very finely punctured and dark brown elytra. The six specimens before me show no variation, but seem to be all females.

71. **Chaloenus aeneipennis**, n. sp. — Flavous the vertex, antennae (the 9th and 10th joint excepted) the tibiae and tarsi black, thorax bifoveolate, elytra greenish-aeneous, deeply punctured in double rows.

Length 2 lines.

Head rather elongate, flavous, the vertex piceous, frontal tubercles strongly raised, lower portion of face forming a single piece, the clypeus scarcely indicated, antennae long and slender, black, the penultimate two joints flavous, basal joint elongate and the longest, slender at the base, thickened at the apex, second joint very short, third, half the length of the first joint, its base fulvous, terminal joints about half the length of the third one but not thickened, thorax transverse, more than twice as broad as long, the sides narrowed at the base, the angles flattened and produced, each with a long seta, the surface with a small fovea at each side, impunctate, the disc stained with piceous, scutellum black, elytra with a transverse depression below the base, with three double rows of deep punctures, the outer ones of which are only visible at the middle where they are bounded at each side by an indistinct ridge, the punctures near the suture are irregular and partly interrupted, underside and legs flavous, tibiae partly or entirely black as well as the tarsi, unarmed, the first joint of the posterior tarsi as long as the following three joints together, claws appendiculate, anterior coxal cavities closed.

Hab. Si-Rambé.

The systematic position of *Chaloenus* is still one of doubt, Chapuis in his " Genera des Coléoptères " placing it amongst the *Halticinae* near *Oxygona,* while Baly considers the proper place to be near *Coelomera* amongst the *Galerucinae.* If the principal character assigned to the *Halticinae,* that is, the thickened posterior femora is to be considered of any value, *Chaloenus* certainly cannot find a place amongst them, since there is scarcely any difference to be found in the thickness of any of the femora and no more than amongst many genera of *Galerucinae;* the unarmed tibiae is another character very rarely found in the *Halticinae,* on the other hand, the shape of the head and the punctate-striate elytra are more suggestive of the latter Subfamily, but in my opinion *Chaloenus* is a form of transition between the two groups of which we have instances enough in other tribes of the animal kingdom. In the present species the apical joints of the antennae are again different from those of the type in having elongate, not thickened joints and the metatarsus of the posterior legs are also much more elongate, but I see no reason to separate the species generically from its allies.

72. **Chaloenus capitatus**, n. sp. — Fulvous, the intermediate joints of the antennae black, apical joints flavous, thorax impunctate, elytra with a few deep punctures at the base only, the disc with a slight purplish gloss.

Mas. Head very broad, the clypeus flattened, black.

Var. Head, thorax and legs nearly black.

Length 2 $1/2$-3 lines.

Head very broad, nearly subquadrate, frontal tubercles small, eyes not large but protruding, clypeus concave, separated laterally by deep oblique grooves, black as well as the sides of the face and the mandibles, labrum fulvous, antennae extending beyond the middle of the elytra, black, the basal three joints fulvous, shining, the apical three flavous, terminal joint black at the apex, basal joint very long, club-shaped, second, short, third slightly shorter than the first joint, the two following joints equal, the rest shorter, slightly widened, pubescent, thorax three times broader than long, the anterior angles obliquely thickened, the sides straight, slightly narrowed at the base, the surface impunctate, fulvous, elytra with a deep transverse depression below the base, within which are some deep punctures, indicative of rows, rest of the surface covered with minute piceous dots, fulvous with a slight purplish gloss, elytral margin acutely raised with a row of deep punctures, underside and legs fulvous.

Hab. Pangherang-Pisang.

In the female, the head is much narrower and the antennae are much shorter; a female from Sumatra contained in my collection has the base of the elytra stained with piceous, also similarly coloured legs and in another variety obtained by Dr. Modigliani, the head and thorax are nearly black, but in other respects these specimens agree; the structure of the antennae and the nearly impunctate elytra separates this species from its allies.

73. Sutrea quadrimaculata, n. sp. — Subquadrate-ovate, flavous, antennae (the basal and apical joints excepted) black, thorax transverse, impunctate, elytra extremely minutely punctured, a transverse band at the base and another below the middle, black.

Length 2 lines.

Head impunctate, frontal tubercles nearly obsolete, clypeus broad and convex, narrowed between the antennae, palpi incrassate, flavous, antennae extending to the middle of the elytra, black, the basal four joints fulvous, the apical one yellowish-white, all the joints rather robust, the third and fourth equal, thorax three times as broad as long, the sides straight, the anterior angles broadly oblique, posterior margin slightly sinuate at each side, the surface entirely impunctate, with an oblique groove near the posterior angles, elytra not perceptibly punctured, flavous, the base with a narrow transverse black band, not extending quite to the sutural margin, a similar band is placed at a little distance from the apex, underside and legs flavous, anterior tibiae unarmed, posterior one with a spur, tarsi piceous, anterior coxal cavities open.

Hab. Si-Rambé.

A single specimen of this species, which differs in the pattern of the elytra from any of its allies is contained in this collection, the general shape is broadly subquadrate and the insect resembles in that respect a species of *Chrysomela,* the posterior femora are strongly incrassate.

74. Hyphasis nigripennis, n. sp. — Fulvous, the antennae (the basal and apical joint excepted) black, head and thorax impunctate, tibiae and tarsi fuscous.

Length 3 lines.

Head impunctate, the frontal tubercles flattened, carina acutely raised into a perpendicular ridge, palpi incrassate, fulvous, antennae extending to the middle of the elytra, black, the basal joint piceous, shining, the apical one fulvous, third and fourth equal, intermediate joints slightly widened, thorax three times broader than long, the sides strongly narrowed in front, greatly rounded at the middle, the lateral margin reflexed, the surface impunctate, fulvous, shining, scutellum fulvous, elytra shining, black, microscopically punctured, their epipleurae broad and concave, underside and legs fulvous, tibiae and tarsi darker, the posterior claw-joint swollen, prosternum longitudinally sulcate. *Hab.* Sumatra (my collection).

Allied to *H. piceipennis*, Baly but larger, the antennae with the terminal joint fulvous, the elytra black without a row of punctures at the margins.

75. **Oedionychis padangensis**, n. sp. — Pale testaceous, the labrum, the antennae and the apex of the posterior femora black, thorax with a basal groove, nearly impunctate, elytra with a short row of punctures below the shoulders and near the suture only.

Length 1 ³/₄-2 lines.

Head with some deep punctures near the eyes and one or two others at the middle of the vertex, the frontal elevations strongly raised, trigonate, clypeus broad, convex. labrum black. antennae not extending to the middle of the elytra, black, the basal joint testaceous, terminal joints rather strongly thickened. pubescent, thorax twice as broad as long, with broad flattened sides, the angles acute, dentiform, the lateral margins strongly rounded anteriorly, the base with a narrow transverse groove, the surface impunctate, scutellum short, transverse, elytra with a narrow longitudinal depression within the shoulders, the depression punctured as well as the space close to the suture anteriorly, the rest of the surface nearly impunctate, but with numerous small piceous spots, the sides of the breast and the apex of the femora more or less black.

Hab. Padang.

A small species, distinct on account of the impunctate thorax, its broadly flattened sides and transverse groove, and by the few punctures of the elytra, placed within the longitudinal depression and near the suture. the claw joint of the posterior tarsi, is as usually strongly swollen, the elytral epipleurae are however scarcely concave, as in *Hyphasis*.

76. **Sebaethe affinis**, Jac. — Specimens from Benculen, Pangherang-Pisang, Si-Rambé and D. Surugnan, which agree with those described by me from other parts of Sumatra in the " Notes from the Leyden Museum " Vol. VI. I have now both sexes before me, the male has the anterior tarsi much dilated, the penis is stout and moderately curved ending in a broad rounded apex with a slight point at the middle, the upper portion near the apex consists of two halfs which nearly meet at the middle but leave a narrow groove or division between them; in the type the underside and legs are flavous, in the specimens now before me, they are more or less black but I am not able to find any other differences of any importance.

77. **Sebaethe melanocephala**, n. sp. — Metallic dark blue, the antennae black, thorax flavous, scarcely perceptibly punctured, elytra violaceous blue, strongly and rather closely punctured, abdomen fulvous.

Length 2 lines.

Head black, the vertex bluish, impunctate, frontal tubercles obsolete, carina strongly raised, palpi very strongly incrassate, the apical joint truncate, black, antennae very closely approached at the base, scarcely extending to the middle of the elytra, black, third and fourth joints elongate, equal, following joints much widened, but slightly longer than broad, pubescent, thorax twice as broad as long, widened at the middle, the sides rounded, narrowly margined, the anterior angles obliquely truncate, the basal margin produced at the sides near the posterior angles and to a greater degree at the middle, where it is broadly rounded, the surface very minutely and sparingly punctured, only seen under a strong lens, scutellum black, triangular, elytra metallic blue, very distinctly and rather closely punctured, legs robust, metallic dark blue, the base of the femora more or less piceous, tibiae broadly channelled, the first joint of the posterior tarsi longer than the following joints together; abdomen flavous.

Hab. Pangherang-Pisang.

A single specimen, allied to *S. violaceipennis*, Jac. but differing in the strong punctuation of the elytra and the flavous abdomen.

78. Sebaethe bifasciata, n. sp. — Blackish, the basal joints of the antennae, the head, thorax and the legs (sometimes) obscure fulvous, elytra very finely punctured, flavous, a transverse narrow band at the base, another band below the middle and the lateral margins anteriorly, black.

Length 1½ line.

Head impunctate, fulvous, the frontal elevations distinct, the clypeus with a distinct central ridge, antennae extending to the middle of the elytra, black, the basal three and the last joint fulvous, basal joint long, the second one half the length of the third joint, the latter as long as the fourth, thorax three times broader than long, the sides strongly rounded, narrowly margined, the surface convex, impunctate, scutellum black, elytra finely and closely punctured, the interstices uneven, flavous, the base with a narrow black band, which, at the sides is connected with another band below the middle, this latter band is gradually narrowed towards the suture which it scarcely reaches, posterior tibiae deeply chanelled, the first joint of the posterior tarsi as long as the following three joints together, legs more or less piceous, the anterior ones paler.

Hab. Si-Rambé.

S. bifasciata is closely allied to *S. 4-pustulata*, Baly from Java of which it may possibly be a variety, but in that species, the antennae have no apical fulvous joint, the posterior black band extends across the suture, the latter as well as the apical margin is likewise black, all which is not the case in the present insect of which three specimens are before me.

79. Chaetocnema Modiglianii, n. sp. — Below nearly black, slightly aeneous, above bronze-coloured, the basal joints of the antennae and the legs flavous, posterior femora aeneous, head granulate and finely punctured, thorax strongly punctate, elytra deeply punctate-striate, the interstices costate at the sides.

Length 1 line.

Head finely granulate, closely and distinctly punctured, often

with a narrow central smooth space, frontal elevations absent,
lower portion finely pubescent, antennae scarcely extending to
the middle of the elytra, black, the lower three or four joints
flavous, third and fourth joints equal, slightly longer than the
second, but much more slender, thorax scarcely twice as broad
as long, the sides nearly straight, the anterior angles oblique,
but to a small extent, the surface rather closely and strongly
punctured at the sides, more remotely so at the middle, where
there is generally a short longitudinal space, free from punctures, basal margin not accompanied by an impressed line or
groove, elytra pointed at the apex, the punctures much larger
than those of the thorax, closely placed, the first interspace at
the suture more irregularly punctured, those at the sides longitudinally costate, the punctures at the apex rather smaller
than those at the base, but very distinct, elytral epipleurae with
a few deep punctures only, posterior femora aeneous, penis
slightly curved, of even width, its apex obliquely rounded, its
extreme apex slightly truncate.

Hab. Padang.

To be separated from other Malayan species by the closely
punctured head, the strong punctures of the thorax and the still
more strongly punctured elytra; there is however some difference
amongst the specimens before me, in regard to the punctuation
of the head, which is in some rather strongly and evenly and
in others more finely punctured, with a smooth narrow central
space, more or less distinct, the thorax also shows some depression near the base in some specimens, but I am not able to
separate these forms satisfactorily.

80. **Chaetocnema granulicollis**, n. sp. - Below nearly black,
the basal joints of the antennae and the tibiae and tarsi, flavous,
thorax finely granulate and punctured, greenish, elytra blackish-
green or brownish, strongly punctate-striate, the interstices
costate.

Length ³/₄ line.

Head very finely granulate and extremely finely punctured,
only visible under a very strong lens, obliquely grooved above

the eyes, the antennae rather widely separated by the transversely quadrate clypeus which is bounded above by a shallow groove, labrum piceous, antennae not extending to the middle of the elytra, black, the lower five joints flavous, second joint as thick as the first, the three following ones short and equal, terminal joints slightly thickened, thorax twice as broad as long, the sides perfectly straight, the anterior angles slightly oblique, the basal margin broadly but gradually produced towards the middle, the surface minutely granulate, scarcely more distinctly punctured than the head, the punctures rather closely placed, the colour greenish, of silken appearance, but little shining, scutellum transverse, small, elytra brownish-aeneous, strongly punctate-striate, the punctures closely placed, the interstices longitudinally costate, underside piceous, the posterior femora greenish, the extreme apex of the femora and the tibiae and tarsi flavous, the posterior tibiae strongly toothed before the apex.

Hab. Siboga.

This small species may be distinguished by the minutely granulate head and thorax and the very fine punctuation of the latter, that of the head being scarcely perceptible; whether a single specimen with much more distinct costate elytra and extremely close punctuation, represents the female of the present species I am unable to say, but it is probable, since all other characters agree; there are four specimens before me. The species can scarcely be distinguished from some Central American forms notably *C. parcepunctata*, Crotch, except in the want of a short row of punctures at each side near the basal margin of the thorax.

81. **Chaetocnema sumatrana**, n. sp. — Below dark fulvous, above piceous, the basal joints of the antennae and the tarsi paler, head impunctate, thorax minutely punctured and granulate, elytra deeply and regularly punctate-striate, the interstices longitudinally costate and minutely punctured.

Length 1 $\frac{1}{2}$-1 $\frac{3}{4}$ line.

Head entirely impunctate, finely granulate, frontal tubercles indistinct, carina broad and flat, clypeus not raised, clothed with some pubescence, labrum fulvous, palpi paler, eyes widely separated, large, antennae extending to the middle of the elytra, fulvous, the apical joints fuscous, the third and following joints nearly equal, rather thick, pubescent, terminal joint more elongate, pointed, thorax twice as broad as long, the sides straight, the anterior angles oblique, the basal margin truncately produced at the middle, not accompanied by a transverse groove or line, the surface finely but not very closely punctured, the interstices minutely granulate, scutellum narrowly transverse, elytra not depressed below the base, strongly and very regularly punctate-striate, the punctures closely placed, the interstices strongly longitudinally costate and very finely punctured, underside and legs dark brown, impunctate, posterior tibiae with a strong tooth, tarsi more or less flavous.

Hab. Pangherang-Pisang.

A comparatively large-sized species, which may be separated by the entirely impunctate head, the costate and finely punctured elytral interstices and the fulvous underside and legs, the colour of the upper surface also differs from so many other species of the genus in having no metallic gloss.

82. **Chaetocnema tenimberensis**, Jac. — The species described by me under the above name in " Novit. Zoolog." 1894, Vol. I. ought, I think, to find its place in *Crepidodera* near *C. costipennis*, Baly to which it seems closely allied. It has also been obtained by Dr. Modigliani at Pangherang-Pisang.

83. **Chaetocnema placida**, n. sp. — Dark bronze-coloured, the basal joints of the antennae and the legs fulvous, head closely punctured and finely granulate, thorax strongly and subremotely punctured, with a central basal smooth space, elytra deeply punctate-striate, the interstices smooth, posterior femora stained .with aeneous.

Length $\frac{1}{2}$ line.

Of subquadrate, posteriorly distinctly narrowed shape, the head finely granulate, rather strongly and closely punctured, with a more or less well marked central longitudinal smooth space, frontal tubercles not visible, clypeus coarsely punctured, its anterior margin nearly straight, furnished with short greyish hairs, antennae nearly extending to the middle of the elytra, black, the lower four or five joints fulvous, second joint slightly thickened, as long as the third, fourth and following joints slightly longer, thorax twice as broad as long, of equal width, the sides straight at the base, the anterior angles obliquely cut, posterior margin slightly rounded, not accompanied by a groove, the surface more strongly but not so closely punctured than the head, the interstices shining, the base with a short central longitudinal smooth space, scutellum broader than long, elytra with deep round punctures, rather widely separated, as well as the rows themselves, the interstices impunctate, smooth and shining, those at the sides slightly costate, the punctures distinct to the apex, legs fulvous, the posterior femora more or less aeneous, underside purplish, distinctly punctured, prosternum rugosely punctured.

Hab. Balighe, Padang, Pea Ragia.

Allied to *C. malayana*, Baly but smaller, the head not rugose-punctate, the thorax without reticulate lateral interspaces, the elytral punctuation not approached in pairs and the interstices smooth and shining.

84. **Chaetocnema Gestroi**, Jac. — Specimens from Pangherang-Pisang and Siboga ; I have also received this species from Borneo, it was originally obtained by Dr. Modigliani at the neighbouring island of Nias, and may be known by its black shining colour and the nearly impunctate thorax.

85. **Sphaeroderma limbatipennis**, n. sp. — Fulvous, the terminal joints of the antennae and the labrum black, thorax finely and closely punctured, elytra closely and irregularly punctate, blackish, the lateral and apical margins more or less broadly fulvous.

Length 1 $\frac{1}{4}$-1 $\frac{1}{2}$ line.

Head impunctate, shining, fulvous, the frontal tubercles small, but strongly raised, carina short, but acute, labrum black, palpi flavous, the penultimate joint swollen, antennae extending nearly

to the middle of the elytra, black, the lower five joints fulvous,
third and fourth joints equal, fifth slightly longer, the following
ones slightly thickened but elongate, thorax of usual shape,
strongly transverse, the sides nearly straight, the anterior angles
rather obtuse, posterior margin strongly bisinuate at each side,
the surface very closely and finely punctured, slightly more
strongly so near the base, scutellum small, triangular, elytra
scarcely more strongly punctured than the thorax, the punctures closely and irregularly placed, rather more regularly arranged in rows at the sides, the punctures extending nearly to
the lateral margins, the disc nearly black, the sides and apex
more or less fulvous, underside and legs fulvous, the breast
more or less black, prosternum transversely quadrate, nearly
impunctate.

Hab. Pangherang-Pisang.

In the black labrum and the colour of the elytra, this species
differs from any of its allies.

86. **Sphaeroderma Wallacei**, n. sp. — Flavous, the breast black,
thorax extremely minutely punctured, elytra metallic blue, finely
and closely punctured, the apex flavous.

Length 1 line.

Head impunctate, the frontal elevations flattened and indistinct,
the carina distinct, labrum fulvous, antennae entirely fulvous,
pubescent, the second and third joints equal, short, the others
gradually but slightly, thickened, thorax about twice and a half
broader than long, the sides straight, the anterior angles obliquely thickened, the surface microscopically finely punctured,
posterior margin oblique, scarcely sinuate at the sides, scutellum black, elytra slightly wider at the base than the thorax,
bright metallic blue, distinctly and closely punctured in irregular
rows, the interstices somewhat wrinkled and also minutely punctate, the apex flavous, underside and legs fulvous, breast black,
clothed with yellow pubescence.

Hab. Si-Rambé.

This species is again closely allied to several other Eastern
forms, with neither of which I can identify it, it differs from

S. cyanipennis, Jac. in the entirely fulvous antennae and the flavous apex of the elytra, from *S. apicalis*, Baly in the bright blue not black elytra and the closely and extremely minutely punctured thorax, from *S. apicipennis*, Baly in the fulvous labrum and antennae as well as in the blue colour of the elytra, in *S. Beccarii*, Jac. likewise from Sumatra, the thorax is more distinctly punctured and the elytra are blackish and the breast fulvous.

Weise has separated certain species from Malacca and Sumatra from *Argopus* (Insecten Deutschlands) and proposed the name of *Dimar* for those in which the antennae are widely separated and the structure of the head is said to differ. I fail however to see the necessity for the erection of another genus and these Eastern species may well be included in *Sphaeroderma*. In *Argopus Ahrensi* for example, the antennae are quite as widely separated and intermediate degrees amongst the exotic species are frequent in almost all structural details, so that a new genus might be equally well established on almost every species. nothing is constant or gained in separating species on slight differences and erect continually new genera on variable structural characters.

87. Sphaeroderma flavoplagiata, n. sp. — Reddish-fulvous, antennae (the basal joints excepted) black, thorax very finely and subremotely punctured, elytra closely and semiregularly punctured, black, the extreme apex fulvous, a large ovate patch on the disc, flavous.

Length 1 ½ line.

Head impunctate, the frontal elevations acutely raised, narrow; clypeus deflexed, labrum and mandibles prominent, penultimate joint of the palpi, incrassate, antennae extending to the middle of the elytra, black, the lower three joints and the extreme apex of the terminal one, fulvous, third joint slightly shorter than the second, following joints rather robust, pubescent, much longer than broad, thorax rather more than twice as broad as long, widened at the middle, the sides nearly straight, anterior margins obliquely thickened, posterior margin with the median lobe broadly rounded, the surface scarcely perceptibly and not very closely punctured, reddish-fulvous, scutellum piceous, elytra slightly wider at the base than the thorax. much more strongly punctured than the thorax, the punctures arranged in closely approached semiregular rows, black, almost the entire disc occupied by a large ovate yellow patch, the extreme apex fulvous, underside and legs, fulvous.

Hab. Si-Rambé. A single specimen.

88. **Sphaeroderma nigromaculata**, n. sp. — Flavous, the terminal joints of the antennae and the legs black, thorax and the legs black, thorax and elytra minutely and closely punctured, each elytron with a central black spot.

Length $1\frac{1}{2}$ line.

Head impunctate, flavous, the frontal elevations obsolete, eyes large, terminal joint of the palpi strongly incrassate, antennae very closely approached at the base, not extending to the middle of the elytra, black, the lower four joints flavous, the terminal joint tipped with fulvous, joint three and four, equal, thinner than the second one, terminal joints gradually thickened, thorax more than twice as broad as long, widened at the middle, the sides strongly rounded and narrowed anteriorly, with a narrow reflexed margin, the anterior angles but slightly thickened, rounded, posterior margin strongly sinuate at each side near the scutellum, the surface extremely finely and rather closely punctured, scutellum small, triangular, elytra wider at the base than the thorax, scarcely more strongly punctured than the latter, the punctures arranged in very closely approached but rather irregular rows, a broad space in front of the lateral margin impunctate, the colour flavous, with a small round black spot at the middle of each elytron, their epipleurae broad and concave, underside fulvous, legs nearly black, the base of the femora and the tip of the tibiae slightly stained with fulvous, tibiae longitudinally chanelled, first joint of the posterior tarsi longer than the second, third joint widened, fulvous, prosternum scarcely longer than broad, mesosternum very short.

Hab. Si-Rambé. A single specimen.

Allied in coloration to *S. biplagiata*, Jac. from Burmah, but the thorax with more strongly rounded sides and angles, the surface not wrinkled, the punctuation of the elytra stronger and their sutural and lateral margins not black. I give here the description of another similarly marked species from Java.

89. **Sphaeroderma javanensis**, n. sp. — Testaceous, antennae black, the apical two joints fulvous, thorax and elytra very minutely punctured, the latter with a large central black spot, legs more or less black.

Length 2 $\frac{1}{2}$ lines.

Much larger than the preceding species and of more ovate shape, the head impunctate, the frontal elevations broadly sub-quadrate, distinct, clypeus very broad and flattened, extending upwards between the tubercles and widely separating them, its anterior margin straight, apex of mandibles black, palpi robust, black, their apical joint conical, thorax three times broader than long, of nearly equal width, the sides nearly straight, but slightly narrowed anteriorly, the anterior angles obliquely thickened, the median lobe only slightly produced, the sides with an oblique depression near the posterior angles, the surface with some extremely fine punctures, only visible under a strong magnifying power, scutellum rather broad, triangular, elytra widened below the middle, moderately convex, their surface more distinctly punctured than the thorax, irregularly arranged, the sides and apex nearly impunctate, the middle with a large round black spot, elytral epipleurae broad and flat, the base of the femora and the tibiae and tarsi black, tibiae not chanelled, the first joint of the posterior tarsi elongate, longer than the following two joints together, prosternum slightly longer than broad, mesosternum short.

Hab. Java (my collection).

This species differs in structural details entirely from the preceding one and would perhaps justify the erection of another genus, but at present more material is needed for the proper classification of these variable *Halticidae* before a proper definition can be arrived at.

90. **Sphaeroderma affinis**, n. sp. — Fulvous, the terminal joints
of the antennae black, thorax distinctly but rather remotely
punctured, elytra black, rather strongly punctured in semiregular
not very closely approached rows.

Length 1 line.

Head impunctate, the frontal tubercles small, but distinct,
carina short, palpi not thickened, flavous, antennae nearly extending to the middle of the elytra, black, the lower six joints
flavous, terminal joints thickened, thorax twice as broad as long,
widened at the middle, the sides slightly rounded, the anterior
angles thickened, the surface finely and rather remotely punctured, the punctures rather more closely placed and stronger
at the base, scutellum black, elytra somewhat oblong, pointed
at the apex, much more strongly punctured than the thorax,
the punctuation rather regularly arranged in not very closely
approached rows, underside and legs fulvous or flavous.

Hab. Pangherang-Pisang and Pea-Ragia.

This species has the same coloration as several other Eastern
forms, but is smaller than any of them, and is distinguished
either by differing in coloration or sculpture of one or other
parts. *S. Beccarii,* Jac. is larger and more closely punctured;
S. Balyi, Jac. has a black or piceous underside and geminate
punctate-striate elytra; in one specimen, an obscure fulvous spot
is visible at the middle of each elytron, there are three specimens before me.

91. **Sphaeroderma sumatrana**, n. sp. — Fulvous, the antennae
(the basal four joints excepted) black, the apical joints widened,
thorax very finely and closely punctured, elytra finely punctured
in rows, the interstices minutely punctate, breast piceous.

Length 1 line.

Head impunctate, eyes very large, the frontal elevations small,
subquadrate, carina acuta, lower portion of face produced, deflexed, antennae extending to the base of the elytra only, the
lower three or four joints flavous, the rest black, widened, pubescent, third and fourth joints equal, as long as the second,
thorax more than twice as broad as long, narrowed in front,

the sides nearly straight, slightly oblique near the anterior angles, the basal margin broadly produced at the middle, surface very minutely and rather closely punctured, elytra ovate, very finely punctured in rows, the interstices scarcely perceptibly punctured, underside and legs fulvous, breast more or less piceous, finely punctured.

Hab. Pangherang-Pisang, also Perack.

Closely allied to several other Eastern forms especially to *S. fascicornis*, Baly from Japan but differing in the straight, not rounded sides of the thorax and in the elytral punctuation being arranged in rows, the antennae also have dilated and pubescent terminal joints which will further help in the recognition of the species.

92. **Sphaeroderma Modiglianii**, n. sp. — Fulvous, the antennae black, three basal joints fulvous, thorax microscopically punctured, elytra very finely and regularly punctate-striate.

Var. Below black.

Length 1 ¼ line.

Head impunctate, antennae robust, extending to the middle of the elytra, the three or sometimes four basal joints fulvous, the others black, pubescent, the second, third and fourth joints very nearly equal, subquadrate, the following ones longer, terminal joint sometimes fulvous, thorax strongly widened at the middle, the sides rounded, narrowly margined, anterior angles scarcely thickened, medial lobe broadly rounded, the surface extremely finely and closely punctured, scutellum very small, elytra very finely punctured in rather closely approached rows.

Hab. Si-Rambé.

Larger than *S. sumatrana* and the rows of elytral punctures more widely separated; these species of *Sphaeroderma* are however very difficult to separate owing to the great variability, scarcely two specimens being alike, the structure of the antennae varying similarly as well as the elytral punctuation, and many specimens are necessary for examination.

93. **Sphaeroderma Rafflesi**, n. sp. — Flavous, above fulvous, antennae black, the basal three or four and the apical joint,

flavous, thorax very finely and closely punctured; elytra distinctly and semiregularly punctured in very closely approached longitudinal rows.

Length 1 ½ line.

Head impunctate, frontal tubercles broad and flat, labrum large, flavous, apex of mandibles black, antennae robust, extending to the base of the elytra, the second and the two following joints equal, short, terminal joints thickened, longer than broad, pubescent, the lower four joints and the apical one flavous, the others black, thorax nearly three times broader than long, the sides straight at the base, rounded near the apex and slightly narrowed at the same place, with a narrow reflexed lateral margin, the anterior angles rounded, the basal margin sinuate at each side, the median lobe broadly rounded, the surface extremely finely and closely punctured, elytra more strongly punctured than the thorax, the punctures arranged in very closely approached not very regular rows, the broad space in front of the lateral margin impunctate, anterior penultimate joint of the tarsi broadly dilated in the male, the last abdominal segment incised at each side, the median lobe broadly subquadrate with a central longitudinal groove.

Hab. Pangherang-Pisang, Sumatra (my collection) also Perak.

Of this species I possess three specimens, of which one has the last joint of the antennae and the sides of the elytra from the base to the middle, black, but does not differ in any other way; the species is larger than the preceding two and differs in several details, notably in the broad and flattened frontal elevations of the head, which in the allied forms are small and subquadrate, also more distinctly raised, a single specimen from Pangherang-Pisang as well as one from Perak in my collection has the breast and the legs more or less black, but does not seem to differ in other respects. *S. antennata*, Jac. from Birmah is closely allied but has longer antennae, the frontal tubercles are acutely raised and the elytral punctuation at the sides arranged in several distinct rows.

94. **Sphaeroderma tibialis**, n. sp. — Fulvous, the antennae

(the basal joints excepted) and the tibiae and tarsi black, thorax nearly impunctate, elytra very finely and closely punctured, black.

Length 2 lines.

Head impunctate, fulvous, frontal tubercles trigonate, flattened, carina rather broad, antennae extending to the base of the elytra, black, the lower three joints flavous, terminal joints widened, pubescent, thorax twice and a half broader than long, the sides straight, much narrowed towards the apex, the anterior angles thickened, the surface scarcely perceptibly and closely punctured, scutellum fulvous, elytra black, finely punctured in closely approached and irregular rows which near the lateral margin are arranged in pairs, elytral epipleurae fulvous, the anterior portion of the breast and of the first abdominal segment margined with black, femora fulvous, tibiae and tarsi black.

Hab. Sumatra (my collection).

Smaller than *S. flavicollis,* Jac. the antennae and legs differently coloured, the sides of the thorax straight etc.

95. **Eucycla nigrofasciata**, n. sp. — Rounded, convex, testaceous, the antennae (the basal joints excepted) the breast and legs black, thorax sparingly punctured, elytra regularly punctate-striate, a transverse band at the base and another near the apex, black.

Length 1 1/4 line.

Head rather long, impunctate, testaceous, the frontal elevations scarcely indicated, the clypeus in the shape of an elongate triangle forming a ridge at each side which extends to the base of the antennae, labrum piceous, palpi slender, antennae extremely closely approached at the base, extending beyond the middle of the elytra, black, the basal three and the last joint flavous, first joint very long and slender, second joint short, third, one half longer, fourth, as long as the two preceding joints together, the following ones slightly widened, pubescent, eyes rather elongate, thorax nearly three times broader than long, the sides nearly straight, obliquely narrowed towards the apex, the anterior angles obliquely thickened, the posterior mar-

gin obliquely sinuate at each side, the surface very finely but
sparingly punctured, flavous, scutellum small, black, elytra convex, with about ten rows of distinct punctures, flavous, the base
occupied by a narrow black band which widenes towards the
suture, down which it extends nearly to the middle, at the lateral margins it is connected with another transverse band below
the middle which is widened at the sides and does not quite
extend to the suture, breast and legs black, abdomen flavous,
posterior tibiae armed with a short spine, prosternum broader
than long, subquadrate, mesosternum scarcely visible, anterior
coxal cavities open.

Hab. Si-Rambé.

There are two specimens of this species contained in the present collection for which Baly's genus *Eucycla* seems to be the
best place; the structure of the clypeus is rather peculiar and
the antennae are very closely approached at the base, their long
basal joint and the punctate-striate elytra agree with the typical
forms.

96. **Eucycla limbatipennis,** n. sp. — Subhaemispherical, piceous,
the basal joint of the antennae and the apical segment of the
abdomen, fulvous, thorax strongly transverse, finely punctured,
elytra punctate-striate, the interstices finely punctured, flavous,
the suture narrowly, and the sides more broadly piceous.

Length 1 1/2 line.

Head with a few minute punctures, black, the frontal tubercles
only indicated by narrow transverse ridges, clypeus raised into
a triangular acute ridge, labrum and palpi obscure fulvous, eyes
very large, elongate, antennae closely approached at the base,
extending to the base of the elytra, the basal joint elongate
and slender, stained with fulvous, the following two joints short,
equal, fourth and following joints more elongate and slightly
thickened, black, thorax nearly three times broader than long,
the sides straight, narrowed in front, the anterior angles
thickened, not produced, the posterior margin broadly rounded
and produced at the middle, the surface finely, not very closely
and irregularly punctured, scutellum small, elytra very convex

and rounded, finely punctured in distant rows, the interstices rather closely and nearly as strongly punctured, flavous, the suture, the sides and the apex piceous, the dark lateral portion nearly impunctate, underside and legs nearly black, prosternum strongly transverse, mesosternum not visible, metasternum broad, its apex broadly rounded, strongly punctured, tibiae somewhat prismatic, triangularly depressed, carinate, the posterior ones mucronate, the first joint of the posterior tarsi as long as the following two joints, the anterior coxal cavities open.

Hab. Si-Rambé. A single specimen.

The punctured striae of the elytra are not very strong and partly obsolete on account of the almost equally strongly punctured interstices, the structure of the head, the long basal joint of the antennae and the strongly transverse prosternum will prevent this species being mistaken for a *Sphaeroderma*.

97. **Eucycla elegantula**, n. sp. — Rounded, fulvous, antennae with the fifth to the tenth joint black, thorax rather finely and closely punctured, elytra strongly punctate-striate, metallic dark blue, the extreme apex fulvous.

Length 1 ½ line.

Head impunctate, without frontal elevations, fulvous, clypeus rather concave with lateral and central ridge, eyes large, antennae closely approached at the base, the lower three joints and the apex of the terminal one, fulvous, the others black, first joint very long and curved, its base very slender, the apex thickened, second joint short, third slightly longer, the following joints thickened, cylindrical and pubescent, apical joint thinner, thorax three times broader than long, the sides nearly straight, narrowed towards the apex, anterior angles obliquely thickened, posterior margin sinuate at each side, rather sharply produced at the middle, the surface finely and rather closely punctured, fulvous, elytra regularly and rather strongly punctate-striate, the striae widely placed, the interstices flat and impunctate, their epipleurae very broad, concave, underside and legs fulvous, the posterior femora very strongly thickened, short, their tibiae dilated, channelled, with a spur, the first joint of the posterior tarsi scarcely longer than the second one, prosternum strongly transverse and short, metasternum strongly punctured, anterior coxal cavities open.

Hab. Sumatra (my collection).

Closely allied to *E. aeneipennis*, Baly but with differently coloured underside and legs and the elytral interstices not convex, the apex also fulvous.

98. **Argopistes flavicornis,** n. sp. Fulvous, the thorax and the breast black, the former finely and rather closely punctured, elytra finely punctate-striate, black, a large oval patch on the disc and the apical margin, fulvous.

Length 1-1 ½ line.

Head fulvous, impunctate, the frontal elevations only indicated in the shape of two very narrow oblique ridges above the antennae, clypeus triangularly raised, fulvous as well as the labrum, palpi slender, antennae very closely approached at the base, fulvous, scarcely extending to the base of the thorax, the basal joint very long and slender, the second one short, thickened, the following four joints very small, nearly equal, the terminal five joints much thickened, forming an elongate club, thorax twice as broad as long, strongly widened at the middle, narrowed in front, the sides nearly straight, the anterior angles slightly thickened, posterior margin strongly sinuate at the sides, the surface finely and rather closely punctured, black, scutellum small, black, elytra closely and distinctly punctate-striate, underside and legs fulvous, breast black, posterior femora strongly incrassate, tibiae broad and flattened, subprismatic, the first joint of the posterior tarsi slightly longer than the second one, prosternum narrowly transverse, mesosternum not visible, the anterior coxal cavities open.

Hab. Pangherang-Pisang.

The two specimens before me vary greatly in size but not in any other essential point, they have the appearance of a small species of *Coccinella* like the other representatives of the genus and almost the same coloration as *A. biplagiatus*, Motsch. and *A. insularis*, Jac., but the punctuation of the elytra differs from the first named species and the coloration of the antennae. legs and underside, from that of the last.

99. **Argopistes Beccarii**, n. sp. — Fulvous. thorax very finely punctate, elytra finely and closely punctured with irregular rows of larger punctures, the middle of each with a round black spot. Length 2 lines.

Of strongly rounded, convex shape, the head impunctate, the eyes rather closely approached, frontal tubercles small, palpi distinctly incrassate, antennae entirely fulvous, rather short, the basal joint elongate, the second short and thick, the following two joints thinner and shorter, the others slightly widened, thorax strongly curved, much widened at the middle, nearly three times broader than long, the sides rounded, the posterior margin strongly oblique at the sides and slightly sinuate ; the median lobe short and truncate, the surface very finely and rather closely punctured, scutellum triangular, elytra convex, wider at the base than the thorax, very finely punctured, the punctures mixed with larger ones, which near the sides are here and there arranged in irregular longitudinal rows, their epipleurae placed deeply inwards, posterior femora strongly incrassate, tibiae widened anteriorly, deeply chanelled, tarsi broad, the first joint of the posterior tarsi, longer than the second one, prosternum longer than broad, narrowed at the middle, the last abdominal segment of the male, with a broadly produced median lobe, the other segments with transverse rows of punctures.

Hab. Mt. Singalang (O. Beccari). A single specimen.

100. **Psylliodes caerulipes**, n. sp. — Below black, above metallic blue, the basal two joints of the antennae flavous, thorax strongly and subremotely punctured ; elytra strongly punctate-striate, the interstices impunctate, costate at the sides. legs metallic blue.

Length 1-1 $\frac{1}{4}$ line.

Of rather broadly ovate shape, the head with a few fine punctures at the vertex, the latter metallic blue, frontal tubercles and carina obsolete, labrum and palpi piceous, antennae extending to the middle of the elytra, black, the lower two joints

flavous, the lower three joints of equal length, the terminal
ones gradually thickened, thorax twice as broad as long, the
sides straight, the anterior angles oblique, forming an angle
before the middle, the surface very distinctly but not very closely punctured, metallic blue, scutellum small, elytra widened
towards the middle, strongly punctate-striate, the punctures
closely placed, the interstices flat, except at the sides, where
they are convex, posterior femora very strongly incrassate, their
tibiae strongly widened at the apex, the latter deeply emarginate below the insertion of the tarsi, the first joint of the
latter longer than the following joints together, abdomen black.

Hab. Si-Rambé.

Differing from *P. Batyi*, Jac. in the unicolorous legs and in
the less closely punctured thorax, also in the more ovate and
less elongate shape. Three specimens were obtained.

101. **Psylliodes fulvipes**, n. sp. — Below piceous, above dark
blue, three basal joints of the antennae and the legs fulvous,
thorax closely and strongly punctured, elytra deeply punctate-
striate, the interstices minutely punctured.

Length 1 1/2 line.

Of comparatively broad shape, the vertex of the head with a
few minute punctures and a small fovea between the eyes at
the lower portion, frontal tubercles nearly obsolete, clypeus
black, mandibles and palpi fulvous, antennae extending nearly
to the middle of the elytra, the lower three joints flavous or
fulvous, the others black, rather robust, third joint distinctly
shorter than the preceding or following one, terminal joints
widened, thorax scarcely twice as broad as long, the sides
straight, the anterior angles oblique, the surface convex, strongly
and closely punctured, scutellum small, black, elytra with closely placed and strong punctures, forming regular rows, the
interstices very finely punctured, legs entirely fulvous, posterior
tibiae strongly widened towards the apex, their upper edge irregular dentate here and there, first joint of the posterior tarsi
longer than the following joints.

Hab. Si-Rambé, D. Surugnan.

Although this species seems very closely allied to *P. Chapuisi,*
Baly it does not agree in several particulars with that species;
it is of broader and more robust shape, the antennae have the
basal three not five joints fulvous, the thorax is more closely
punctured and the legs are entirely bright fulvous; there are
five specimens before me which show no differences in regard
to the above particulars.

102. **Psylliodes baligheensis,** n. sp. — Below piceous, above
metallic blue or greenish, the basal three joints of the antennae
and the four anterior legs (more or less) fulvous, thorax rather
finely and subremotely punctured, elytra finely punctate-striate,
the interstices flat and extremely finely punctured.

Length 1 ¼ line.

Head with a few minute punctures, the eyes large, the frontal
elevations and the carina very small but distinct, labrum and
palpi nearly black, antennae not quite extending to the middle
of the elytra, black, the lower three joints fulvous, first and
second joint equal, third, slightly shorter, terminal joints thickened, thorax about one half broader than long, distinctly narrowed in front, the sides straight, the anterior angles oblique,
posterior margin rather strongly rounded and produced at the
middle, the surface finely and not very closely punctured, elytra
finely punctate-striate, the interstices not costate, here and there
impressed with a few minute punctures, posterior femora more
or less metallic blue, tinged with fulvous, their tibiae of the
latter colour or piceous, without teeth along the upper edges,
abdomen obscure fulvous.

Hab. Balighe, October and March.

Of this species, two specimens are contained in this collection,
they seem to me to differ from their allies in the finer and
more remote punctuation of the thorax as well as of the elytra
which have also non-costate interstices at the sides, the locality
in which these specimens were obtained also differs from that
of others. *P. Balyi,* Jac. of which numerous specimens are also
contained in the present collection, is generally metallic green
above, the thorax is very closely and almost rugosely punctured

and the elytral punctuation is strong and close as well as the interstices convex.

103. **Psylliodes nigroaenea**, n. sp. — Blackish-green, basal joints of the antennae, fulvous, legs more or less piceous, thorax finely subrugosely punctured, elytra strongly punctate-striate, the interstices scarcely raised and punctured.

Length ³₄-1 line.

Head impunctate, the frontal tubercles small and obsolete, palpi fulvous, antennae not extending to the middle of the elytra, black, the lower four joints fulvous, the second joint as long as the first, the following two joints slightly shorter, the terminal joints thickened; thorax nearly twice as broad as long, the sides straight, the anterior angles oblique, the basal margin scarcely produced at the middle, the disc rather convex, closely punctured and somewhat rugose or wrinkled with traces of longitudinal striae here and there, scutellum small, elytra strongly punctate-striate, the punctures very closely placed, forming striae at the sides, the interstices at the same place slightly convex, legs more or less piceous, tibiae and tarsi paler, the posterior tibiae without teeth excepting the one near the apex.

Hab. Si-Rambé.

This is a species of comparatively short and broad shape, of an almost black colour with a slight aeneous gloss, it is much less elongate than *P. Balyi*, Jac. and of entirely different coloration, the thorax also is of nearly equal width and not narrowed in front which will assist in the recognition of the present species of which I have four specimens for comparison.

104. **Allomorpha glabrata**, n. sp. — Black, the base of the antennae and the tibiae and tarsi fulvous, above greenish-blue, opaque, entirely impunctate, non-pubescent.

Length 1 line.

Similar in shape to a species of *Aphthona*, the head impunctate, the vertex bluish-green, the lower portion blackish, the frontal tubercles in the shape of narrow oblique ridges, clypeus with a longitudinal groove down the centre, labrum and palpi black, penultimate joint of the latter slightly thickened, apical joint acutely pointed, antennae long and slender, black, the lower four joints fulvous, second joint proportionately thick and long, but slightly shorter than the third joint, fourth, longer than the preceding one, the following joints much more elongate and slender, thorax subquadrate, scarcely one half broader than long, the sides nearly straight, the angles not produced, the surface with a very obsolete transverse sulcation before the middle, entirely glabrous and impunctate, greenish-blue, opaque, scutellum blackish, elytra of the same colour and sculpture as that of the thorax, their epipleurae broad at the base, the underside black, femora with a bluish tint, the posterior ones very strongly incrassate, tibiae fulvous, the posterior ones dilated towards the apex, not chanelled, with a distinct spine, anterior tibiae unarmed, tarsi fulvous, the metatarsus of the posterior legs as long as the following three joints together, claws appendiculate, prosternum narrow, the anterior coxal cavities closed.

Hab. Pangherang-Pisang and Siboga.

A. glabrata differs from the other two species contained in this genus, in the absence of any pubescence of the thorax and elytra, but agrees in the want of any visible punctuation and other particulars with the typical forms; all are of delicate and silky appearance as only found amongst the *Halticidae* in the bladder-clawed *Monoplatinae*; I know however only of three species.

GALERUCINAE

105. **Aulacophora similis**, Oliv. — Several specimens of this common and widely distributed species were obtained at Siboga, Balighe, Pea Ragia and Pangherang-Pisang. The variety *A.* (*Rhaphidopalpa*) *flavipes*, Jac. is also amongst them. In the Deutsche Entomolog. Zeitschrift for 1892, Weise has changed the name of *Aulacophora* and substituted several other genera, on account of structural characters differing. I must regret both these alterations, the name of *Aulacophora* has been used ever since 1834 and although a similar genus exists in Botany, no great con-

fusion can in my opinion arise since everyone knows whether in a description, the Animal or Vegetable kingdom is meant and changes of names once familiar, do not add to the assistance of the memory, which in Entomology is already more than too heavily taxed. The other genera created by Weise are principally founded on male structural characters which vary almost with every species so that about another fifty or more genera might easily be made; Weise has moreover taken no notice of Baly's papers on the genus *Aulacophora* (Journ. Linn. Soc. V. XX, 1886 and 1888) where most of the species are characterised and tabulated in a useful way and which must have the priority in regard to specific names.

106. Pseudocophora sumatrana, n. sp. – Below black, above flavous as well as the anterior legs, sulcus of thorax straight, elytra strongly and semi-regularly punctured.

♂. Elytra with a basal fovea, the latter bidentate, anteriorly, last abdominal segment trilobate, its median lobe nearly flat.

Length 3 lines.

Head impunctate, flavous, frontal tubercles narrowly transverse, clypeus with an acute ridge, palpi incrassate, antennae only extending to the base of the elytra, pale flavous, the third joint as long as the first and longer than the following joints, thorax twice as broad as long, the sides straight at the base, rather strongly rounded anteriorly, the surface entirely impunctate, the basal sulcus straight and deep, elytra strongly punctured in closely approached very irregular rows, their epipleurae continued to the apex, undersides and the four posterior legs black, claws bifid.

Hab. Si-Rambé.

From *P. brunnea*, Baly and *P. nitens*, All. the present species is at once distinguished by the black underside and similar coloured posterior legs, it resembles in this coloration several true species of *Aulacophora* from which it can readily be separated by the prolonged elytral epipleurae, the female has, as usual simple elytra and entire last abdominal segment.

107. Phyllobrotica elegantula, n. sp. — Flavous, the antennae (the basal joints excepted) the apex of the tibiae and the tarsi, fuscous. thorax transversely sulcate, impunctate, elytra finely pubescent, extremely closely punctured. fuscous, the base narrowly fulvous.

Var. Elytra flavous, the sides more or less fuscous. Length 2 lines.

Narrowly elongate. the head broad and convex behind the eyes. fulvous shining, impunctate, eyes rather large, frontal tubercles trigonate, clypeus triangularly raised, labrum nearly black, palpi flavous, the penultimate joint incrassate, antennae long and slender, fuscous, the basal three joints flavous, the first joint rather long and curved, the second very short, the third nearly double as long, fourth joint slightly longer than the following joints, thorax narrowly transverse. the disc transversely sulcate, impunctate, shining, flavous, elytra very finely and closely punctured, clothed with short whitish pubescence, the basal portion more or less flavous, the rest fuscous, elytral epipleurae scarcely visible at the base only, legs slender, flavous, tibiae unarmed, the first joint of the posterior tarsi longer than the following joints together, claws appendiculate, anterior coxal cavities open.

Hab. Si-Rambé.

Closely allied to *P. javana*, Jac. but differing in the black labrum, the colour of the elytra and in the longer metatarsus of the posterior legs, in one specimen the scutellum is black.

108. **Mimastra sumatrensis**, Jac.   I refer specimens obtained at Pangherang-Pisang to this species (Notes Leyd. Mus. Vol. VI). I have not now the type before me, but my description agrees in everything except in that of the length of the antennae, which in the type, is given as "nearly as long as the body"; in the present insects the antennae in the male extend considerably beyond the elytra, those of the females slightly so; the elytra are very finely and closely punctured but scarcely rugose as in the type and it is therefore possible that I have another closely allied species before me, the entire insect is of a uniform pale testaceous colour.

109. **Hoplasoma frontalis**, n. sp. — Below black, the head, antennae and thorax testaceous, elytra metallic dark green, closely rugose-punctate, tibiae and tarsi black, claws bifid.
Length 3 lines.

Head impunctate, the eyes very large, frontal elevations trigonate, contiguous, clypeus broadly triangular extending upwards between the antennae and forming a single, scarcely depressed piece, palpi rather long and slender, antennae very long and slender, nearly as long as the body, testaceous, the third joint smaller than the fourth, the second one very small, thorax one half broader than long, the sides straight, the surface transversely sulcate at the middle, impunctate, pale testaceous as well as the scutellum, elytra elongate, dark metallic green, rugosely punctured throughout, their epipleurae rather broad at the base but disappearing at the middle, legs long and slender, tibiae unarmed, the first joint of the posterior tarsi as long as the following three joints together, femora more or less testaceous, their apex and the tibiae and tarsi black.

Hab. Si-Rambé.

Differing in the structure of the clypeus and (*H. metallica*, Jac. excepted) in the metallic green and rugose elytra ; their epipleurae are broader at the base than in the typical forms but other differences of importance are not present.

Sosibiella, gen. n.

Body elongate, antennae filiform, the second and third joint short, head broad, thorax transverse, sulcate, elytra glabrous, irregularly punctured, their epipleurae rather broad, obsolete below the middle, legs slender, tibiae unarmed, the metatarsus of the posterior legs as long as the following two joints together, claws appendiculate, anterior coxal cavities open.

This genus will have to be placed near *Haplosomoides*, Duviv. from which it differs in the swollen head, the shorter third joint of the antennae and in the transverse thorax.

110. **Sosibiella caeruleipennis**, n. sp. — Flavous, the head and

thorax impunctate, elytra metallic blue, closely and strongly punctured, the interstices irregularly rugose, the tibiae and tarsi black.

Length 3¼ lines.

Head broad and convex at the vertex, impunctate, eyes small, the frontal elevations trigonate, small, clypeus broad, depressed at the apex, not well separated, labrum and palpi flavous, the latter not incrassate, antennae two-thirds the length of the insect, flavous, the second joint slightly smaller than the third, both joints short, the fourth joint longer than the fifth, pubescent like the rest of the joints, the terminal two joints thin and shorter, thorax short and transverse, the sides nearly straight and slightly narrowed at the base, the angles slightly thickened, the disc broadly transversely sulcate at the middle, scutellum broad, flavous, elytra wider at the base than the thorax, closely and strongly punctured, the interstices rugose and wrinkled, underside and the femora flavous, finely pubescent, the apex of the femora above and the outer and inner margins of the tibiae as well as the tarsi, black.

Hab. Pangherang-Pisang. Two specimens.

111. **Microlepta fulvicollis**, n. sp. — Flavous, tibiae black, base of head and the thorax fulvous, the latter finely punctured, elytra closely and finely punctured, the interstices slightly rugose.

Length 2¼ lines.

Head impunctate, the vertex fulvous, the frontal tubercles trigonate, rather small, lower portion of face forming a single piece, flavous, eyes large, palpi elongate, slender, antennae extending to the end of the elytra, filiform, the basal joint extremely long, the second one very short, the following joints very long and slender, finely pubescent, lower two joints dark fulvous or piceous, thorax nearly twice as broad as long, of equal width, the sides slightly rounded at the middle, rather suddenly constricted near the anterior angles, the latter obsolete, slightly thickened, posterior angles acute, the surface finely, irregularly and closely punctured with some obsolete depressions at the sides anteriorly, dark fulvous, the anterior margin more or less narrowly flavous, scutellum triangular, elytra wider at the base than the thorax, flavous, finely and closely punctured in very irregular rows, the interstices finely wrinkled, their epipleurae broad at the base, indistinct below the middle, legs slender, tibiae mucronate, blackish, the first joint of the posterior tarsi, much larger than the following joints together, claws appendiculate, anterior coxal cavities open.

Hab. Pangherang-Pisang.

This species agrees in every structural details with the other ones forming the genus *Microlepta*, notably in the long slender antennae and their very long basal joint, five other species from the Malayan regions have been described by myself of which *M. pallida* is the nearest allied one, but differs in the uniform colour of the legs and other details.

112. **Pseudoscelida apicicornis**, n. sp. — Elongate, black, the apical four joints of the antennae, flavous, head and thorax flavous, the latter transversely foveolate, impunctate, elytra metallic dark blue or greenish, very closely and strongly punctured.

♂. Antennae with a long fringe of hairs.

Length 3 lines.

♂. Head broader than long, the vertex impunctate with a triangular fovea, eyes very large, frontal elevations very narrow and obsolete, clypeus triangular, its apex extending in shape of an acute ridge between the antennae, labrum and palpi, black, antennae long and slender, black, shining, the last four joints fulvous or flavous, each joint furnished with a fringe of long hairs, the first joint long and slender, the second extremely small, the following joints equal, the fifth and sixth, rather curved, terminal joints smaller, thorax one half broader than long, of equal width, rather deflexed near the anterior angles, the sides nearly straight, the angles not produced, the surface with a transverse depression at each side, impunctate, shining, scutellum broad, triangular, black, elytra very closely and strongly impressed with round punctures, their epipleurae rather broad, continued below the middle, legs slender, black as well as the underside, tibiae unarmed, the first joint of the posterior tarsi as long as the following joints together, claws appendiculate, the anterior coxal cavities open.

Hab. Pangherang-Pisang.

This is the second species of the genus described by me in " Novitat. Zoolog. " 1894; it agrees in all characters with the type and the male has likewise the fringe of long hairs attached to the antennae, which in the female is wanting, the general appearance of the insect is not unlike that of a *Mimastra* but the thorax is differently shaped and foveolate instead of sulcate, the antennae also have their joints of different proportionate length.

113. **Sastra sulcicollis,** n. sp. — Elongate, obscure pale fuscous, pubescent, thorax impunctate, the posterior half transversely sulcate, elytra minutely punctured and clothed with fine fulvous pubescence.

Length 3 $\frac{1}{2}$ lines.

Head impunctate, the frontal tubercles distinct and oblique, clypeus deflexed. smooth. impunctate. eyes prominent and rounded, antennae slender, filiform. extending to nearly two-thirds the length of the elytra, pale flavous, the third joint very long, more than twice the length of the second one, thorax twice as broad as long, the sides slightly rounded before the middle, rather suddenly narrowed anteriorly, the anterior angles tuberculiform, the disc shining, deeply transversely sulcate posteriorly, the depression bounded anteriorly and laterally by a strongly raised ridge, posterior margin rather broadly produced at the middle, elytra clothed with fine adpressed fulvous hairs. extremely finely and closely punctured, legs slender. tibiae unarmed, the first joint of the posterior tarsi, longer than the following joints, claws bifid, the posterior ones much less acutely divided than the anterior ones, anterior cavities open.

Hab. Si-Rambé, Pangherang-Pisang.

The general colour of this species is a dull and opaque pale fuscous, the structure of the thorax will at once distinguish it from any other contained in this genus, principally distinguished by the long third joint of the antennae.

114. Cerophysa Gestroi, n. sp. — Metallic violaceous-blue, thorax bifoveolate, impunctate, elytra very closely and strongly punctured.

♂. Antennae with the eighth joint elongate and strongly dilated.

♀. Antennae with simple, short joints, the apical two flavous. Length 3 lines.

Head impunctate, metallic blue, frontal elevations entirely absent, clypeus strongly raised in shape of a triangular ridge, palpi robust, the apical joint conical, antennae extending to about the middle of the elytra, the second joint very short, the third longer than the following four joints which are triangularly widened, the eighth elongate and greatly dilated, cylindrical, the terminal three joints short again; thorax one half broader than long, the sides straight at the base, rather deflexed anteriorly where the angles are obsolete, the surface with a deep fovea at each side, extremely finely and sparingly punctured, only visible under a strong lens, scutellum broad, blackish, elytra without basal depression, extremely closely and rather strongly and somewhat rugosely punctured, the basal portion more finely punctured, below metallic blue, as well as the legs, tibiae unarmed, the first joint of the posterior tarsi as long as the following three joints together, anterior coxal cavities open.

Hab. Pangherang-Pisang.

In shape and colour, the present species exactly resembles the type *C. nodicornis,* Wied. but the antennae differ totally in structure, having the dilatation transferred to only one joint, the eighth; this may perhaps be considered sufficient by some, to place the insect in another genus, but in my opinion, structures like this, peculiar to the antennae in the male sex only are extremely variable and do not justify the erection of new genera when other characters of distinction are absent. *C. borneoensis,* Jac. agrees very nearly with the present species but has entirely differently structured antennae, the female of that species is more difficult however to separate from that of the species before us and can only be distinguished by the more elongate joints of the antennae, those of *C. Gestroi* are very short and gradually thickened towards the apex, that is if the specimens before me are really the females of *C. Gestroi*. Of *C. borneoensis*, a single male specimen from Si-Rambé was likewise obtained by Dr. Modigliani.

115. **Luperus sumatranus**, n. sp. — Black, finely pubescent, the basal joint of the antennae and the femora more or less fulvous, thorax transversely sulcate, elytra extremely finely punctured and pubescent, black.

Length 1 ¹/₄ line.

Of narrow and parallel shape, the head broad, impunctate, the frontal elevations scarcely defined, eyes very large, occupying the entire sides of the head, clypeus raised in shape of a narrow ridge, palpi fulvous, antennae long and slender, black, the first joint fulvous, long and stout, the second one very short, the following joints as long as the first, pubescent, thorax subquadrate, one half broader than long, the sides straight, the surface impunctate, transversely sulcate, shining, elytra very minutely and rather closely punctured, clothed with single whitish erect hairs, below black, the four anterior or all the femora fulvous, the first joint of the posterior tarsi as long as the following three joints together, claws appendiculate.

Hab. Si-Rambé.

The species somewhat resembles *L. hirsutus*, Jac. from China but is much smaller and has differently coloured femora.

116. **Luperodes** (*Cneeodes*) **bisignatus**, Motsch. — The Sumatran specimens, obtained at Pea Ragia, Lago Toba, Balighe, and Dol. Tarabugua etc. agree entirely with Motschulsky's description of his type from Birmah, but as Weise has rightly remarked, the species cannot find its place in *Monolepta* where it is placed in the Munich Catalogue, on account of the open coxal cavities, but must be incorporated with *Luperodes;* the Sumatran specimens are frequently devoid of the dark elytral spot, this variety has been described by myself under the name of *Luperodes scutellatus* (Notes Leyd. Mus., 1884). Whether *Monolepta bimaculata*, Hornst. is specifically distinct is somewhat doubtful ; the male

of the present insect has a small oblong fovea below the scutellum on each elytron.

117. Galerucella sumatrana, n. sp. — Obscure pale testaceous, pubescent, the intermediate joints of the antennae, black, thorax finely rugose, obsoletely depressed at the disc, elytra finely pubescent, the base and a lateral stripe, obscure fuscous.

Length 3 lines.

Of nearly parallel shape, the head finely but distinctly punctured, rather shining, eyes large, frontal elevations distinct, rather broad, clypeus triangularly raised, palpi slender, the terminal joint acutely pointed, antennae extending to the middle of the elytra, the basal two and the apical four joints testaceous, the others black, the third joint one half longer than the second but half the length of the fourth joint, the latter the longest, thorax twice and a half as broad as long, of equal width, the sides straight from the base to the middle, from thence to the apex, narrowed, the angles distinct but not produced or dentiform, the surface with a shallow depression at each side, finely rugosely punctured and clothed with pale pubescence, posterior margin obliquely shaped at the angles, scutellum broad, pubescent, elytra broader at the base than the thorax, finely granulate and pubescent, the shoulders rather prominent, the sides deflexed, bounded above by an obsolete ridge, the base and a broadish stripe from the shoulders downwards, very obscure fuscous, epipleurae narrow, extending nearly to the apex, legs slender, unarmed, the metatarsus of the posterior legs as long as the following two joints together, claws bifid, anterior coxal cavities open.

Hab. Pangherang-Pisang.

This species may be known by the coloration of the antennae and that of the elytra, the general colour is a very pale testaceous and the darker markings are very obscure, they may be more plainly visible in other specimens but I have only two examples before me.

118. Malaxia Weisei, n. sp. — Narrowly elongate, black, head finely rugose, thorax with deep lateral depression, black, elytra metallic green, finely granulose, femora fulvous, tibiae and tarsi black.

Length 2 3/4-3 lines.

Head finely rugose, black, opaque, the middle with a central groove, frontal elevations small, shining, clypeus swollen, transverse, shining, palpi black, antennae extending to about the middle of the elytra, black, pubescent, the basal joint slender, curved, shining, its base flavous, second joint one half shorter than the third, fourth joint twice as long as the third, terminal joints shorter, thorax twice as broad as long, the sides strongly rounded, the surface very finely rugose and clothed with fine yellowish pubescence, the anterior margin with a rather deep indentation at the middle, the disc with a deep and broad depression at each side, black, scutellum black, elytra bright green, finely rugose, the extreme lateral margins purplish, the surface clothed with very fine yellowish pubescence; femora fulvous, tibiae and tarsi black, claws bifid.

Hab. Si-Rambé.

Amongst the many nearly similarly coloured species of *Malaxia* now known, the present one can only be compared with *M. nigricollis,* Alld. from Africa and *M. assamensis,* Jac.; it is however much larger than either of those species and differs in the colour of the antennae and legs and other details, amongst the species from the Malayan regions, no other is known to me which has an entirely black head and thorax. I have much pleasure in dedicating this species to my friend and collegue Herr Weise in Berlin.

119. **Atysa (?) frontalis,** n. sp. — Fulvous, finely pubescent, antennae (the lower three joints excepted) black, thorax finely granulate, with lateral depressions and central groove, elytra closely and distinctly punctured, clothed with short fulvous hairs.

Length 2 lines.

Of elongate and nearly parallel shape, the head rather produced and elongate, the vertex finely rugose, the lower portion of the face shining, forming a single piece, divided at the middle by a longitudinal ridge, which extends upwards between the

antennae, palpi incrassate, fulvous, antennae extending to the
middle of the elytra, black, the lower three joints fulvous,
the first and third, the longest, equal, terminal joints shorter,
slightly thickened, pubescent, thorax nearly three times broader
than long, the sides rounded at the middle, the anterior angles
in shape of small tubercles placed a little below the anterior
margin, posterior margin, sinuate at the middle, the surface
broadly but shallowy depressed at each side, closely covered
with short pubescence, obscuring any punctuation, fulvous, scutellum not longer than broad, its apex truncate, the surface finely pubescent, elytra distinctly and very closely punctured,
slightly rugose, clothed with short fulvous pubescence, their
epipleurae broad, continued to the apex, legs rather robust, the
posterior femora rather thickened, tibiae unarmed, not sulcate,
the first joint of the posterior tarsi as long as the following two
joints together, claws bifid, prosternum narrow, but distinct,
anterior coxal cavities open.

Hab. Si-Rambé.

I have for the present included this species in *Atysa* from
which it only differs in the structure of the head, which is peculiar for a *Galerucide*, although not unique (*Coelocrania*) in
Atysa the frontal tubercles, clypeus etc. are well separated and
distinct and do not form one piece as is the case here. The insect is of an entirely fulvous and opaque colour with the exception of the antennae.

120. **Atysa dimidialipennis**, n. sp. — Black, the head and thorax
fulvous, finely granulate and pubescent, elytra sculptured like
the thorax, the anterior half fulvous, the rest black, opaque.

Length 3 lines.

Elongate, slightly widened posteriorly, head broad, finely rugose and pubescent with a narrow central groove, frontal tubercles trigonate, smooth, shining, divided by a deep groove, clypeus
in shape of a highly raised transverse narrow ridge, deflexed
anteriorly, labrum fulvous, antennae black, pubescent, the second
joint small, the third and fourth equal, dilated and robust as
well as the fourth joint, which is slightly smaller, all the other

joints broken off, thorax twice as broad as long, not widened at the middle, the sides straight, the disc with a deep central longitudinal groove, the sides with a more obsolete transverse depression, the surface finely rugose and pubescent, fulvous, elytra more finely rugose or granulate than the thorax, clothed with similar pubescence, with three very obsolete longitudinal costae, only visible in certain lights, the anterior half fulvous, the posterior one black, the line of division of the two colours not sharply defined, slightly oblique, underside and legs black, the first joint of the posterior tarsi as long as the following two joints together, claws bifid.

Hab. Si-Rambé. A single specimen.

In the coloration of the elytra, this species agrees partly with *A. terminata*, Baly but in that insect, the head and thorax are black and the fulvous portion of the elytra occupies two-third of their length.

121. Atysa imitans, n. sp. — Below and the legs black, above fulvous, the head with a black spot, thorax finely pubescent with central groove and lateral depressions, elytra minutely punctured, fulvous, pubescent, the apex obscure fuscous.

Length 3 lines.

Of very narrow shape, the head finely granulate and pubescent, opaque, the vertex with a large blackish spot, the frontal tubercles very small, clypeus rather broad, with a distinct central ridge, labrum blackish, antennae scarcely extending to the middle of the elytra, black, the third and fourth joints equal, the following gradually shorter, terminal joint thinner, thorax twice as broad as long, the sides very slightly rounded at the middle, the surface closely and finely pubescent, fulvous, with a deep longitudinal central groove of dark colour, the sides with a deep oblique depression extending nearly to the margins, scutellum truncate at the apex, pubescent, elytra minutely punctured and pubescent like the thorax, of the same colour, the apex obscure fuscous at the sides and near the suture, indistinctly separated from the fulvous portion, the apical half with an obsolete longitudinal raised costa, at the middle, extending upwards towards the middle of the elytra, the first joint of the posterior tarsi as long as the following two joints.

Hab. Si-Rambé. A single specimen.

Of half the width of the preceding species and although closely allied, apparently distinct in the shape of the antennae, the black vertex of the head and the lateral deep fovea of the thorax which makes the anterior portion appear to be raised.

122. **Xenoda basalis**, Jac. (The Entomologist Suppl. 1893). — A small and extremely variable species in regard to the colour of the elytra which are either black or testaceous or partly of either colour, and clothed with fine pubescence, the antennae are strongly swollen in the male but have no spine as in the type *X. spinicornis,* Baly, but in the female they are long and slender. Specimens were obtained at Si-Rambé and Pangherang-Pisang. I may add here, that the tibiae in *Xenoda* are unarmed and that the anterior coxal cavities are open.

123. **Xenoda pallida**, n. sp. — Pale testaceous, finely pubescent, thorax transversely sulcate, impunctate, elytra very finely rugosely punctate and pubescent.

♂. Antennae with the intermediate joints greatly dilated, furnished with a long spine.

Length 3 lines.

Head impunctate, frontal tubercles strongly raised, trigonate, clypeus in shape of a transverse ridge, palpi robust, antennae scarcely extending to the middle of the elytra, pale testaceous, the intermediate joints very strongly and gradually widened, the eighth joint furnished with a long acute spine, ninth joint very elongate, the terminal two shorter, thorax very short and transverse, the disc transversely sulcate, impunctate, elytra finely rugose and sparingly clothed with pale pubescence, tibiae sometimes fuscous, the first joint of the posterior tarsi longer than the following three joints together.

Hab. Si-Rambé.

X. pallida is evidently closely allied to *X. hirtipennis,* Jac. also from Sumatra, it is of the same size and general colour and has similarly constructed antennae, but differs in the finely rugose elytra and their uniform pale coloration. I have not now the type of X. *hirtipennis* before me but as I have stated particularly in my description, that the elytra of that species are *not* rugose I must look upon the present insect as distinct; there are three specimens before me.

124. **Xenoda nigricollis**, n. sp. — Black, the apical two joints of the antennae flavous, elytra rather strongly rugose, aeneous, purplish or blue.

♂. Antennae with the intermediate joints widened and swollen, furnished with a long spine.

Length 3 lines.

This species agrees in all structural characters with X. *spinicornis*, Baly but the thorax is entirely black and the elytra are more strongly rugose and very sparingly pubescent, the underside and legs are black, the antennae have the intermediate joints greatly dilated and widened as in the typical form; the female has, as usual, simple antennae, all the joints, with the exception of the small second one, are of nearly equal length; four specimens are before me which were obtained at Si-Rambé.

125. **Xenoda abdominalis**, n. sp. — Dark purplish, finely pubescent, the antennae black, the apical two joints flavous, thorax impunctate, elytra finely rugose, abdomen flavous.

♂. The intermediate joints of the antennae moderately dilated.

♀. Antennae simple and filiform.

Length 2 lines.

Head impunctate, the vertex convex, frontal tubercles strongly raised, transverse, labrum and palpi (sometimes) flavous, antennae extending beyond the middle of the elytra, black, the apex of the ninth joint and the apical two joints flavous, the third to the fifth swollen, but longer than broad, the following joints normal, the apical two elongate, thorax of usual shape, short and transverse, sulcate throughout, elytra finely rugose, clothed with black pubescence, tibiae and tarsi blackish, abdomen flavous.

Hab. Si-Rambé.

This species may be at once known by the but moderately dilated antennae and the want of a spine in the male as well as by the colour of the abdomen.

126. **Haplosonyx basalis**, n. sp. — Black, the head, antennae, thorax and the anterior legs fulvous, thorax finely punctured, deeply sulcate, elytra very closely and irregularly punctured, fulvous, the basal portion and the extreme apex black.

Var. Elytra fulvous.

Length 3 lines.

Head impunctate, the frontal tubercles and the clypeus thickened, palpi swollen, antennae extending beyond the middle of the elytra, pale flavous, opaque, the lower three joints shining only, second and third joint very short, fourth longer than the fifth, thorax more than twice as broad as long, the sides straight, the anterior angles oblique, the surface with a deep transverse sulcation at each side, interrupted at the middle, rather finely and sparingly (sometimes more strongly) punctured, scutellum black, elytra very closely and more distinctly punctured than the thorax, the basal portion, extending nearly to the middle, as well as the extreme apex, black, the rest fulvous, the underside and the four posterior legs more or less black, the anterior ones fulvous, sometimes all the tarsi of the latter colour.

Hab. Si-Rambé, Pangherang-Pisang.

Of only half the size than *H. speciosus,* Baly which the present species somewhat resembles in colour; the irregular and close punctuation agrees with *H. nigricollis,* Duviv. but of course, not the coloration of the thorax and upper surface; the variety differs in nothing except in the fulvous colour of the elytra which show however a trace of the black apex.

127. **Cynorta sumatrana**, Jac. — Three specimens obtained at Si-Rambé show some slight differences from the type (Notes Leyden Mus. 1878, Vol. IX) they are larger, the labrum is stained with piceous as well as the tibiae and tarsi; the lower portion of the face is bright flavous and the thorax shows some traces of punctuation, it is therefore possible that I have a closely allied species before me. *C. parvula,* Jac. is smaller and

has a flavous lateral margin of the thorax as well as a continued sulcus across the middle of the latter part.

128. **Cynorta monstrosa**, n. sp. — Below black, the head, three
basal joints of the antennae, the thorax and the legs flavous,
thorax bifoveolate, elytra metallic green, finely punctured and
granulate, the sides with a longitudinal depression.

♂. Head deeply excavated above the clypeus, with a projection
above, the sides of the clypeus as well as the palpi, black, the
latter swollen, antennae with the sixth joint emarginate.

♀. Head and antennae simple.

Length $1\,^1/_2$-2 lines.

♂. Of narrow, elongate shape, the head flavous, impunctate, the frontal tubercles strongly raised, ending in a short
projection above the deep excavation, the space below the eyes
also obliquely excavated and bounded by the lateral ridge of
the clypeus, the sides of the latter black, palpi black, strongly
incrassate at the penultimate joint, antennae two-thirds the
length of the body, fuscous, the lower four joints flavous, basal
joint rather short and robust, second, very short, third and fourth
joint equal, sixth strongly emarginate at its upper edge, ending
into an acute point, terminal joints elongate, thorax one half
broader than long, subquadrate, the sides narrowed at the base,
the angles distinct, with a single seta, the surface impunctate,
flavous, with a broad fovea at each side, scutellum black, elytra
metallic green, minutely punctured and granulate, the sides with
a short longitudinal sulcus which is much more deeply punctured, than the rest of the surface, underside black, legs flavous,
the first joint of the posterior tarsi, longer than the following
joints together, tarsi slightly stained with fuscous, anterior coxal
cavities closed.

Hab. Si-Rambé.

The structure of the head in the male differs from that of
C. capitata, Jac. as well as the antennae; in the female the basal
joint of the antennae is long and slender and the head is simple,
that is to say, the clypeus is triangularly raised at the margins,
its surface being depressed or sulcate and the palpi are less

strongly swollen, in all other respects, this sex agrees with the male.

129. **Anthipha inclusa,** n. sp. - The vertex of the head, the terminal joints of the antennae, the breast, tibiae and tarsi black, thorax flavous, elytra finely and subremotely punctured, flavous, the basal margin, a narrow transverse band before the middle and the apical portion black.

Length 2 lines.

Head impunctate, black, the lower portion flavous, frontal tubercles distinctly raised, oblique, clypeus flattened and rather broad, labrum and palpi piceous, antennae extending to the middle of the elytra, black, the lower five joints flavous, third joint scarcely twice as long as the second one, fourth and fifth equal, elongate, thorax nearly three times broader than long, of equal width, the sides straight, the anterior angles slightly thickened, the disc rather convex, finely but not closely punctured, flavous, scutellum black, elytra very nearly punctured like the thorax, flavous, with a narrow band at the base connected at the sides with another one before the middle and widened at the suture, as well as the posterior third portion, black, below flavous, the breast, tibiae and tarsi black, tibiae unarmed, prosternum very narrow and convex, the anterior coxal cavities closed.

Hab. Pangherang-Pisang.

From any of the other species belonging to this genus, the present one differs in the shape of the thorax which has the anterior margin straight, not concave, in the fine punctuation of the thorax and elytra and in the position of the black bands of the latter; in one specimen the second band is interrupted near the lateral margins and the antennae are entirely flavous.

130. **Anthipha similis,** n. sp. — Entirely testaceous, thorax sparingly punctured, elytra closely and rather strongly punctate, the interstices slightly wrinkled or rugose.

Length 3 $\frac{1}{2}$ lines.

Hab. Si-Rambé.

Whether this species is really distinct from *A. apicipes* or represents only a pale variety I am unable to say with certainty, but I have apparently a male and female specimen before me, which differ in the structure of the antennae from *A. apicipes*. In the male, the second and third joints are short and very nearly equal, the fourth and following joints are opaque and pubescent and very elongate, as long as the basal three joints together, the apex of the terminal joint is black as in *A. apicipes;* in the female, the third joint is also shorter than the corresponding joint in *A. apicipes;* the thorax is similar, but the elytra are more strongly punctured and the interstices more wrinkled, lastly, the entire underside and the legs are flavous and the locality differs also from that of the allied species.

131. **Anthipha variabilis**, Jac. — A single specimen was obtained by Dr. Modigliani (Pangherang-Pisang) which belongs to an entirely black variety in regard to the upper and under surface, the antennae and tarsi alone being pale fulvous in colour, I find that specimens in good condition are clothed with distinct fulvous pubescence at the elytra, which was absent in the specimens which served me for the type. The insect is subject to great variation in regard to colour.

132. **Anthipha cruciata**, n. sp. — Fulvous, thorax short and transverse, impunctate, black, elytra convex, finely and closely punctured, reddish-fulvous, the margins narrowly and a broad transverse band at the middle, black.

Var. a. Black above and below, the elytra with an obscure fulvous spot before and another below the middle.

Var. b. Black, the disc of the elytra fulvous.

Var. c. Fulvous, the tibiae black.

Var. d. Testaceous, the thorax with two, the elytra with five black spots (2.2.1.).

Length 2 ³/₄ lines.

Of ovate and convex shape, extremely variable in regard to coloration, the head impunctate, fulvous, the frontal tubercles narrow, transverse, clypeus triangular, the vertex with a round fovea at the middle, antennae slender, extending to the middle of the elytra, flavous, the basal joint often stained with piceous, the second joint short, the third, half the length of the fourth, following joints of similar length, thorax nearly three times broader than long, the sides slightly rounded, the anterior angles obliquely produced, the disc rather flattened at the middle, with a few fine punctures at the sides, black, the anterior angles fulvous, scutellum rather broad, black, elytra very obsoletely depressed below the base, finely and rather closely punctured in irregular rows, the sides near the apex with a short costa (♀).

Hab. Si-Rambé.

I have taken for the type, the specimen which seems most plainly marked and in which the elytra may also be described as being black with two large fulvous patches, one before, the other below the middle, but scarcely two specimens are alike in coloration and there exist probably other forms differently marked yet, the species may be separated by its strongly convex shape and the want of any elytral depression, from most of its congeners, also by the entirely fulvous antennae, the female is further distinguished by the short elytral costa although the latter is absent in one specimen.

133. **Anthipha tenuimarginata**, n. sp. — Testaceous, the antennae, tibiae and tarsi, black, head and thorax fulvous, the latter very sparingly punctured, elytra very closely and distinctly punctured, very narrowly margined with black.

Length 3 lines.

Of elongate and nearly parallel shape, moderately convex, the head impunctate, fulvous, the frontal tubercles narrowly transverse, clypeus triangularly raised, palpi acutely pointed, piceous, antennae extending beyond the middle of the elytra, black, slender, the third joint twice as long as the second but half the length of the fourth joint, thorax rather more than twice as broad as long, the sides straight, the angles rather acutely pointed outwards, anterior and posterior margins slightly curved, the disc with a few fine punctures at the sides, fulvous, scutellum triangular, testaceous, elytra very closely and distinctly punctured, the punctures arranged in irregular rows with traces of narrow smooth longitudinal lines, all the margins narrowly black, underside black or testaceous, tibiae and tarsi black, the femora fulvous or more or less stained with piceous along their upper margin.

Hab. Si-Rambé.

Very closely allied to *A. subrugosa,* Jac. from Borneo, but of more elongate shape, the head and thorax darker fulvous, the elytra less strongly and more closely punctured. In one specimen before me, the suture is scarcely black, the underside is pale and the femora almost fulvous but the structural details agree in every respect.

134. **Anthipha apicipes,** n. sp. — Flavous, the breast, the apex of the femora and the tibiae and tarsi black, thorax finely and sparingly punctured, elytra strongly and very closely punctured. Length $3\frac{1}{4}$ lines.

Head impunctate, transversely grooved, frontal elevations flat, transverse, clypeus and labrum fulvous, like the rest of the head, antennae slender, extending slightly beyond the middle of the elytra, flavous, the extreme apex of the terminal joint blackish, the third joint elongate, but shorter than the fourth, following joints nearly equal, thorax twice and a half as broad as long, rather flattened, the anterior and lateral margins straight, anterior angles slightly obliquely truncate, posterior ones acute, the disc with a very obsolete lateral depression and a few fine punctures at the sides, scutellum broader than long, elytra distinctly sulcate within the shoulders, the latter rather prominent, the surface closely and strongly punctured in partly regular partly irregular rows, the breast black at the sides, as well as the legs, the abdomen and the basal two-thirds of the femora, fulvous.

Hab. Pangherang-Pisang.

A typical *Anthipha* of comparatively large size and rather flattened shape, principally distinguished by the colour of the breast and legs; in one specimen however, the abdomen is black as well as the breast, the two specimens before me seem to be females on account of the long third joint of the antennae.

135. **Anthipha latefasciata,** n. sp. — Flavous, the antennae and

the tibiae and tarsi, black, head and thorax fulvous, the latter sparingly punctured, elytra strongly punctured in closely approached rows, black, a very broad oval patch at the disc of each flavous.

Length 3 lines.

Head impunctate, flavous or fulvous, frontal tubercles transverse, clypeus triangular, moderately raised, palpi piceous, antennae black, the second joint half the length of the third, fourth joint nearly as long as the preceding joints together, terminal joint wanting, thorax nearly three times broader than long, the sides straight, the posterior and anterior angles slightly prominent and thickened, with the usual setae, the surface very sparingly punctured at the sides, the latter obsoletely depressed, scutellum large, flavous or fulvous, elytra strongly and closely punctured in longitudinal rows, the narrow black band at the base and the similarly coloured apex separated by a large oval yellowish patch, which occupies the entire disc, underside flavous, finely pubescent, legs more or less black.

Hab. Si-Rambé.

The broad discoidal flavous patch of the elytra and their strong and close punctuation will help to distinguish this *Anthipha* from any of its congeners.

136. **Anthiphula modesta**, n. sp. — Testaceous, antennae, tibiae and tarsi blackish, thorax narrowed at the base, impunctate, elytra strongly and closely punctured.

Length 1 $^3/_4$-2 lines.

Elongate, subparallel, the head rather broad, impunctate, eyes large, frontal tubercles strongly raised, broad, transverse, carina strongly raised, lower portion of clypeus rather deflexed, antennae long and slender, extending beyond the middle of the elytra, black, the basal joint obscure fulvous, long and slender, the second joint half the length of the third, fourth and fifth long, equal, thorax subquadrate, narrowed at the base, the sides rather strongly rounded anteriorly, the angles dentiform, the surface rather convex, impunctate, shining, pale fulvous, scutellum broad, elytra distinctly broader than the thorax at the base, strongly and closely punctured in irregular rows, their epipleurae very broad, continued to the apex, legs rather long, tibiae unarmed, the first joint of the posterior tarsi as long as the following two joints together, claws appendiculate, anterior coxal cavities closed.

Hab. Pangherang-Pisang.

The present genus was founded by me on an insect from Birmah, and described in the Genoa Annals 1892. In that species the legs were more robust and the prosternum distinct, both this is not the case in the species from Sumatra but the shape of the thorax and everything else agrees with the type and I have thought it best to place the insect for the present in this genus.

137. **Emalthea subrugosa**, n. sp. — Ovately subquadrate, flavous, the antennae (the basal joints excepted) black, thorax finely and rather closely punctured, elytra metallic green or blue, very strongly and closely punctured, the interstices rugose.

Length 2 $\frac{3}{4}$ lines.

Head very finely and sparingly punctured, antennae extending to the middle of the elytra in the male, blackish, the basal three joints flavous, the fourth joint quite double the length of the third one, thorax three times broader than long, the anterior angles not produced but strongly oblique, furnished with a single seta, placed below the lower angle, sides nearly straight, the surface finely granulate and minutely and irregularly punctured with some small obsolete depressions, scutellum broad, black, elytra with a shallow transverse depression below the base, very coarsely and closely punctured, the interstices rugose, underside and legs flavous, the extreme apex of the tibiae and the tarsi, piceous.

Hab. Si-Rambé.

Of less convex and dilated shape than the following two species, also rather smaller, the tibiae flavous, the antennae not dilated at the intermediate joints as in *E. Balyi*, the elytra very strongly punctured and more strongly rugose, than in the last named species ; the male organ long and slender, curved, the apex strongly obliquely excavated but scarcely widened towards the point.

138. **Emathea Balyi**, n. sp. — Broadly ovate, rufous, antennae (the basal joints excepted) the tibiae and tarsi black, thorax finely and sparingly punctured, elytra metallic green, closely and strongly punctured, the interstices slightly wrinkled.

Length 3 lines.

Head finely strigose at the vertex, transversely grooved between the eyes, the clypeus finely punctured, broad, flattened at the upper portion and nearly contiguous with the frontal tubercles, labrum flavous, antennae not extending to the middle of the elytra, black, the lower two joints fulvous, the second and third joints very small, the intermediate joints slightly widened, thorax nearly three times as broad as long, the sides moderately rounded, the anterior angles flattened and produced, the surface sparingly and finely punctured, rufous with some obscure piceous spots, scutellum broad, trigonate, black, elytra convex, very strongly and closely punctured, metallic dark green, the interstices more or less wrinkled; underside and legs fulvous, the breast finely pubescent, tibiae and tarsi black, the first joint of the posterior tarsi as long as the following two joints together, claws appendiculate, tibiae unarmed, anterior coxal cavities closed.

Hab. Si-Rambé.

Although closely allied to *E. aeneipennis*, Baly, the present species differs in several particulars, notably in the black antennae and tibiae of all the legs, in the colour of the labrum, the structure of the head and in the great difference of the elytral sculpturing and that of the thorax, also in the wrinkled but not punctured elytral interstices.

139. **Emathea fulvicornis**, n. sp. — Fulvous, thorax very sparingly punctured, elytra depressed below the base, convex, dilated posteriorly, metallic green, closely punctured in irregular rows, underside and legs entirely fulvous.

Length 3 lines.

Closely allied to the preceding species but with entirely fulvous antennae and legs, the head finely strigose and very minutely punctured at the vertex, the frontal tubercles strongly raised, transverse, the clypeus also distinctly raised, much narrower than in the preceding species, antennae slender, scarcely extending to the middle of the elytra, pale fulvous, the third joint much shorter than the fourth, thorax of the same shape as the preceding species, sparingly and finely punctured, elytra very convex and widened posteriorly, with a shallow transverse depression below the base, strongly and very closely punctured in very irregular rows, metallic green.

Hab. Pangherang-Pisang.

E. fulvicornis differs from *E. aeneipennis,* Baly in the colour of the antennae and legs, the same differences separate it from the preceding species as well as the shape of the frontal tubercles and that of the clypeus; there seem to be only female specimens before me; the punctuation of the elytra is not nearly so coarse as in that of *E. Balyi* and the interstices are not so distinctly rugose; a specimen from India contained in my collection does not seem to differ in any way from the Sumatran form.

140. Monolepta Modiglianii, n. sp. — Dark brown, antennae black, thorax extremely minutely, elytra very finely and closely punctured, the interstices finely wrinkled.

Length 1 ³/₄ line.

Of ovate, convex shape, the head impunctate, the frontal elevations narrowly transverse, bounded behind by a deep transverse groove, clypeus rather flattened, eyes large, antennae long, black, the basal joint dark brown, second joint small, third one half longer, following joints elongate, pubescent, thorax twice as broad as long, the sides and the anterior margin perfectly straight, posterior margin slightly rounded, the angles oblique, the anterior angles slightly thickened and oblique, the surface extremely finely and obsoletely punctured, scutellum broadly triangular, elytra very finely but more distinctly punctured than the thorax, legs slender, the extreme apex of the tibiae piceous, the first joint of the posterior tarsi very long, anterior coxal cavities closed, elytral epipleurae very broad at the base, indistinct or very narrow below the middle.

Hab. Si-Rambé.

Allied to *M. castanea,* Alld. but differing in the colour of the antennae and that of the underside and legs, the elytral punctuation also is confused and shows no sign of arrangements in rows.

141. Monolepta approximans, n. sp. — Pale testaceous or flavous, antennae long and slender, thorax transversely sulcate at the sides, closely punctured, elytra as closely but more strongly punctured.

Length 1 $^3/_4$-2 lines.

Of convex, posteriorly slightly widened shape, the head impunctate, the eyes very large, occupying the entire sides of the head, frontal tubercles distinct, trigonate, clypeus broad and flattened, antennae closely approached at the base, extending nearly to the apex of the elytra, testaceous, the first joint long and slender, the second and third short, nearly equal, the following joints as long as the first one, thorax twice as broad as long, the sides perfectly straight, the posterior margin but slightly rounded, the disc with a transverse sulcus at each side, scarcely interrupted at the middle and not extending quite to the lateral margin, the surface finely, closely and somewhat rugosely punctured, scutellum small, triangular, smooth, elytra more strongly punctured than the thorax, the punctures more regularly arranged and distinct to the apex, the interstices slightly wrinkled or rugose, elytral epipleurae indistinct below the middle; underside impunctate, legs slender, the posterior tibiae with a long spur, the first joint of the posterior tarsi half the length of the tibiae, anterior cavities closed.

Hab. Si-Rambé.

Several similarly coloured pale species are known from the Malayan region, the present one may be separated by the perfectly straight sides of the thorax in connection with its sulcus and the close and distinct punctuation of its upper surface.

142. Monolepta latefasciata, n. sp. — Head and thorax fulvous, antennae (the basal joints excepted) the breast and the four posterior legs black, elytra extremely minutely punctured, black, the middle with a broad transverse flavous band.

Length 2 lines.

Head broader than long, impunctate, flavous, frontal elevations narrowly transverse, clypeus triangularly swollen, labrum piceous, palpi rather robust, piceous, antennae extending nearly to the middle of the elytra, black, the lower three joints fulvous, the second and third joint short, the basal joint long, the fourth and following joints pubescent, as long as the basal joint, thorax nearly twice as broad as long, the sides rounded, as well as the posterior margin, the surface transversely convex without depressions, finely but not very closely punctured, anterior angles thickened, scutellum longer than broad, black, elytra convex, extremely closely and finely punctured, the black base and apex separated by a broad flavous transverse band extending to the sides, the anterior edge of this band is convex, the posterior one, concave, elytral epipleurae extremely narrow below the middle, the breast and the posterior four legs black, anterior legs obscure fulvous, abdomen flavous, metatarsus of posterior legs very long, anterior coxal cavities closed.

Hab. Si-Rambé.

Closely allied to *M. flavofasciata,* Jac. from Burmah, but the antennae black not fulvous, the thorax punctured, the elytral flavous band much wider and the apex of the elytra black.

143. **Monolepta militaris**, n. sp. — Reddish-fulvous, the head, antennae, thorax and legs flavous, thorax and elytra finely punctured, the latter reddish-fulvous with a narrow transverse black band at the base.

Length 2 lines.

Head impunctate, flavous, the frontal tubercles transverse, labrum and palpi obscure piceous, antennae extending to the middle of the elytra, pale flavous, the terminal two joints black, the second and third joint small, the fourth as long as the basal joint, thorax scarcely twice as broad as long, the sides straight, the anterior angles oblique, furnished with a single seta placed at the middle and within the angles, posterior margin rounded, the surface finely and closely punctured, scutellum black, elytra moderately convex, nearly parallel, as finely punctured at the thorax, the base to less than a third of their length, black, the rest reddish-fulvous, legs flavous, breast obscure piceous as well as the four posterior femora to a greater or smaller degree, the metatarsus of the posterior legs half the length of the tibiae.

Hab. Si-Rambé.

Smaller than *M. basalis,* Jac. and with differently coloured head, antennae, thorax and legs.

144. **Monolepta marginicollis**, n. sp. — Flavous, a spot on the head, the antennae (the basal joints excepted) and the margins of the thorax narrowly black, elytra extremely finely punctured, flavous, margined with black, abdomen and legs flavous, breast black.

Length 1 $^3/_4$ line.

Head impunctate, the vertex with a short central black stripe, frontal elevations very small, clypeus rather broad and flat, antennae very closely approached at the base, black, the lower three joints flavous, third joint very little longer than the second one, following joints equal, thorax twice as broad as long, the sides slightly rounded, with a notch below the anterior angles within which a single seta is placed, the disc without depressions, scarcely perceptibly punctured, flavous, the margins, with the exception of the anterior one, narrowly piceous, scutellum black, elytra convex, scarcely more distinctly punctured than the thorax, all the margins and a spot on the shoulders, black, the disc also with some very faint longitudinal darkish stripes, abdomen and legs flavous, the metatarsus of the posterior legs half the length of the tibia, the breast black.

Hab. Si-Rambé.

In the system of coloration, this species resembles *M. margi-nicollis* and *M. melanocephala* but the head has a black spot, the thorax has no depressions and its lateral margin has a more or less distinct notch below the thickened anterior angles, in both specimens before me the elytra show traces of faint longitudinal stripes, the thorax in this species has not only the sides but also the posterior margin black.

145. **Monolepta obtusa**, n. sp. — Pale fulvous, the terminal joint of the antennae more or less black, thorax transverse, finely punctured, elytra more distinctly and very closely punctured.

Length 1-1 ¹/₄ line.

Head impunctate, frontal tubercles strongly raised, transverse, antennae extending to the middle of the elytra, black, the lower three joints fulvous, the basal joint elongate, the second and third less than half the length of the preceding one, equal, the other elongate, thorax at least twice as broad as long, the sides slightly rounded, narrowly margined, anterior margin straight posterior one slightly produced at the middle, the surface rather convex, extremely finely and closely punctured, elytra more distinctly and extremely closely punctured, the interstices slightly wrinkled at the sides and minutely punctured, the first joint of the posterior tarsi half the length of the tibia.

Hab. Si-Rambé.

This small species has not much to distinguish it from many of its similarly coloured congeners, it may however be known by the colour of the antennae and the close punctuation of its upper surface in connection with its small size, *M. brunneipennis,* Jac. resembles the present species but is larger and has entirely fulvous antennae. Ten specimens are before me.

146. **Monolepta nigromarginata**, n. sp. — Testaceous, the antennae and the breast, black, thorax narrowly black at the sides, elytra scarcely perceptibly punctured, narrowly margined with black.

Length 1 ¹/₂ line.

Head impunctate, obscure pale fulvous or piceous, frontal tubercles acutely raised, clypeus broad, eyes large, antennae extending to the middle of the elytra, black, the lower three joint fulvous, second and third joint small, the following ones elongate, pubescent, thorax one half broader than long, pale testaceous, the lateral margins narrowly black, the sides nearly straight, the disc very obsoletely transverse sulcate, with a few minute punctures, scutellum black, elytra scarcely more distinctly punctured than the thorax, testaceous, all the margins narrowly black, the legs and abdomen testaceous, the underside of the thorax and the breast, black, elytral epipleurae black, broad at the base, very narrow below the middle, the first joint of the posterior tarsi much longer than the following joints together but shorter than half the length of the tibiae.

Hab. Si-Rambé.

A small species, which may be known by the narrow black margins of the thorax and elytra.

147. **Monolepta melanocephala**, n. sp. — Testaceous, the head, antennae, breast and the tibia and tarsi blackish, thorax obsoletely depressed, finely punctured, elytra minutely punctured, very narrowly margined with piceous, the base more broadly so.

Length 1 3/4 line.

Head impunctate black, frontal tubercles strongly raised, eyes large, antennae scarcely extending to the middle of the elytra, black, the lower three joints obscure flavous, second and third joints small, the following ones elongate, terminal joint widened at the middle, thorax subquadrate, nearly twice as broad as long, the sides straight, the posterior margin slightly rounded, the surface with an obsolete depression at the sides. very finely and closely punctured, scutellum black, elytra nearly parallel, moderately convex, very minutely, but rather more distinctly punctured than the thorax, the interstices slightly wrinkled, all the margins narrowly piceous, this colour slightly more extended at the base and surrounding the shoulders, breast black, tibiae and tarsi nearly black.

Hab. Si-Rambé.

From the preceding species, the present one is at once distinguished by the black head and the colour of the tibiae and tarsi, the insect is also slightly larger and less convex and the thorax is somewhat more square-shaped, the base of the elytra is more broadly although obtusely marked with black.

148. **Candezea thoracica**. n. sp. — Pale testaceous. the extreme

basal margin of the elytra, the scutellum and the breast black, thorax finely and closely punctured, the basal margin produced into a tooth, elytra with an oblique fold across the suture below the base, minutely punctured.

Length 2 lines.

Head impunctate, eyes large, frontal tubercles trigonate, clypeus with a central ridge, labrum piceous, palpi slender, antennae two-thirds the length of the elytra, flavous, the last joint and the apex of the preceding two joints black, basal joint long, second one short, third, shorter than the first joint, following ones elongate, equal, thorax twice as broad as long, of nearly equal width, the sides slightly rounded, the surface very closely and finely punctured, slightly wrinkled, the posterior margin produced at the middle into a blunt tooth, hiding partly the scutellum, elytra rather convex, more finely punctured than the thorax, pale testaceous, the basal margin narrowly black, this colour extending downwards at the sides below the shoulders, each elytron is raised below the base into a short oblique ridge or fold near the suture, the latter is likewise raised at the same place, apex of the elytra truncate, their epipleurae broad and continued to the apex, black at the basal portion only, breast black, abdomen and legs testaceous, all the tibiae armed with a spine, the metatarsus of the posterior legs very long.

Hab. D. Tolong. November 1890.

A single specimen of this curiously structured species is contained in this collection, as the insect is glued to a card I am not able to say with certainty whether the coxal cavities are closed or open, but the general appearance and the broad elytral epipleurae agree best with *Candezea;* no istance of a similarly structured thorax amongst the *Phytophaga* has come under my notice, but whether this is a character peculiar to the male only or not I am unable to say.

149. **Candezea impressicollis**, n. sp. — Elongate, pale flavous, thorax transversely sulcate, finely punctured, elytra distinctly and extremely closely punctured.

Length 2 ½ lines.

Head impunctate, eyes very large, the space dividing them narrower than their diameter, clypeus broad, forming a single piece, antennae inserted near the upper portion of the eyes, closely approached, pale flavous, two thirds the length of the elytra, the first joint long, and slender at the base, the second and third joints short, equal, following joints very elongate and slender, thorax twice as broad as long, the sides perfectly straight, the anterior angles slightly thickened, obliquely shaped, anterior margin straight, posterior one rounded, the disc finely punctured with a transverse distinct depression, not extending to the sides, elytra distinctly, very closely and irregularly punctured, with some finer punctures on the interstices, their epipleurae prolonged below the middle but narrow except at the base, metatarsus of the posterior legs long, anterior coxal cavities closed.

Hab. Pangherang-Pisang.

This is another species of similar uniform coloration as so many other Eastern forms, the comparative large size, shape of the clypeus, straight sides of the thorax and its sulcation will however help to distinguish it; Weise has separated some species with similarly structured head and placed them in his genus *Pseudocrania* but the same structure to a slightly less degree is found in *Candezea* taking *C. cisteloides,* Har. (*occipitalis,* Reiche) for the type, so there is nothing to separate these two genera except the state of the anterior coxal cavities which Weise described as " subobelusa " in *Candezea* they are certainly closed but there is no certainty in regard to this structure as in some instances one side may be found to be slightly open and the other quite closed (in *Ochralea* for instance) the insertion of the antennae at the upper portion of the eyes in opposition to those in *Monolepta* (*M. pauperata* as type) may still be useful in separating *Candezea* from the last named genus ; *Pseudocrania africana,* Jac. (*nigricornis,* Weise) has certainly closed cavities in the specimens I have examined, so that the genus cannot be separated from *Candezea.* The penis in the present insect is short and straight, and gradually narrowed to a point.

150. **Neolepta multicolorata**, n. sp. — Fulvous, antennae robust, black, thorax finely punctured, elytra closely and distinctly punctured, black, the base and the apex fulvous, the disc with a yellowish transverse spot.

Length 2 lines.

Head longer than broad, fulvous, impunctate, the frontal tubercles strongly raised, trigonate, clypeus broad, narrowed into a raised ridge between the antennae, labrum flavous, palpi piceous, not incrassate, antennae rather widely separated, inserted near the lower portion of the eyes, robust, pubescent, the second and third joint very short, the following ones rather widened, elongate, the last two joints wanting; thorax nearly twice as broad as long, the sides slightly rounded as well as the posterior margin, anterior angles obsolete, the surface finely but not very closely punctured, without impressions, scutellum triangular, elytra more strongly and closely punctured than the thorax, black, the basal margin narrowly, the apex more broadly fulvous, the middle with a transverse yellowish spot, underside and legs fulvous, the tibiae and tarsi blackish, the metatarsus of the posterior legs as long as the tibiae, claws appendiculate, anterior coxal cavities closed.

Hab. Sumatra (my collection).

This species, of which I possess a single specimen is closely allied to *N. biplagiata,* Jac. likewise from Sumatra, but differs in the want of depressions at the thorax, the more strongly and closely punctured elytra and their fulvous base and apex; the third joint of the antennae is slightly smaller than the second one (probably the male) and the other joints are very robust.

151. **Macrima costatipennis**, n. sp. — Elongate, testaceous, the breast black, thorax deeply bifoveolate, impunctate, elytra strongly longitudinally costate, the interstices nearly impunctate.

Length 3-4 lines.

♂. Head impunctate, the frontal tubercles transverse, clypeus triangularly raised, palpi robust at the penultimate joint, antennae extending nearly to the apex of the elytra, flavous, the second joint extremely short, the third and following joints very long, equal, basal joint curved and thickened at the apex, thorax about one half broader than long, the sides straight and narrowed at the base, slightly but rather suddenly widened in front, the angles distinct, anterior and posterior margins straight, the disc deeply bifoveolate at the middle, impunctate, scutellum broader than long, elytra parallel, elongate, strongly costate throughout, each elytron with seven or eight costae, the interspaces not perceptibly punctured, legs slender, the first joint of the posterior tarsi much longer than the following joints together, claws appendiculate, anterior coxal cavities closed.

Hab. Si-Rambé.

Several other species, placed in this genus, have been described by myself from the Malayan region, the type was founded by Baly on a species from India; the genus comprises species resembling those of *Aulacophora* but the claws are appendiculate, the elytral epipleurae continued and the anterior cavities are closed; the present species resembles slightly *M. subcostata*, Jac. likewise from Sumatra but the elytra in the insect before me are much more strongly costate and the interstices are scarcely punctured but show small piceous spots; in the specimen which seems to be the male and which is smaller than the females, the second joint of the antennae is extremely small, the corresponding joint in the other sex is rather larger, I am only able to discover a spine at the intermediate tibia in the male, but in the females, all the tibiae have a distinct long spine, otherwise there seem to be no differences between the two sexes.

152. **Theopea Modiglianii**, n. sp. — Reddish-fulvous, the lower joints of the antennae, the apex of the tibiae and the tarsi more or less black, thorax subquadrate, bifoveolate, elytra violaceous blue, closely punctured and transversely rugulose.

Length 2 $\frac{1}{2}$-3 lines.

Head impunctate, the frontal elevations strongly marked, trigonate, palpi piceous, antennae long and slender, pubescent, black, the basal joint fulvous at the base, rather slender, the second one very short, third and following joint of nearly equal length, the last four joints yellowish white, thorax not longer

than broad, subquadrate, shining, impunctate, bifoveolate at the
disc, elytra closely and irregularly punctured, the interstices
transversely and irregularly rugose throughout, without costae;
penis long and slender, rather suddenly and strongly widened
near the apex, the latter produced into a short hook-like point.
Hab. Si-Rambé.

Of this species there are two male specimens before me which,
like several other closely allied species have the same colouration
as the common *T. impressa*, Fab. *T. Modiglianii* is however at
once distinguished from the latter species by the absence of the
intermediate dilated joints of the antennae and the want of costae
of the elytra; the penis is also of different structure, that of
T. impressa, although equally slender, tapers gradually towards
the apex instead of being widened at that portion, and ends
into a rather acute triangular point without tooth or hook.

153. **Theopea clypeata**, n. sp. — Fulvous, the antennae black,
the apical three joints yellowish-white, thorax biimpressed, with
purplish gloss, elytra metallic blue, closely and strongly punctured, the interstices obsoletely costate at the sides ($\female$).

Length 2 lines.

Head minutely granulate, impunctate, the vertex with a purplish tint, the clypeus in shape of an acutely raised triangular
or semicircular ridge and connected with another acutely raised
ridge dividing the antennae, palpi piceous, antennae slender,
pubescent, black, the apical three joints whitish, third joint
slightly shorter than the fourth, thorax not longer than broad,
of usual shape, slightly narrowed at the base, the disc bifoveolate at the middle, finely granulate and very minutely punctured,
scutellum fulvous, elytra strongly and irregularly punctured, the
interstices unevenly rugose and obsoletely longitudinally costate
towards the sides, underside and legs fulvous, the metatarsus
of the posterior legs very long, tibiae unarmed, coxal cavities
closed.

Hab. Pangherang-Pisang.

There are two female specimens only contained in this collection. they have more the appearance of a *Cynorta* than that

of one belonging to the present genus, but the pubescent antennae, unarmed tibiae etc. show the true characters of *Theopea;* the species is smaller than most of its congeners and differs in the structure of the clypeus, the irregular punctured elytra and their obsolete costae from *T. impressa* and other allied species; the male differs probably again in having differently structured head and antennae, the head and thorax in the present species show also a metallic hue which is absent in any of the others.

154. **Hyphaenia apicicornis**, n. sp. — Testaceous, the basal joints of the antennae and the tibiae and tarsi more or less piceous, thorax subquadrate, bifoveolate, elytra metallic blue or green, semi-punctate-striate, the interstices longitudinally costate.

♂. Antennae longer than the body, with long fringes of hairs.

♀. Antennae shorter, without long hairs, the elytra geminately punctured.

Length 2-3 lines.

Of narrow, elongate shape, the head impunctate, testaceous, the frontal elevations transverse, eyes large, clypeus in shape of a narrow transverse and perpendicular ridge, penultimate joint of the palpi, incrassate, antennae longer than the body, very slender and elongate, the first joint thickened, glabrous, the second very small, the following joints very long and slender, all furnished with long fringes of hairs, the lower six joints black, the others flavous, thorax subquadrate, one half broader than long, the sides straight, slightly narrowed before the middle, the angles not produced, the disc transversely sulcate, impunctate, flavous, scutellum broad, flavous, elytra broader at the base than the thorax, parallel, distinctly punctured in semiregular rows, the punctures of transverse shape, the interstices finely granulate and closely longitudinally costate, below and the legs testaceous or flavous, tibiae unarmed, the first joint of the posterior tarsi longer than the following joints together, claws appendiculate, anterior coxal cavities closed.

Hab. Si-Rambé.

The size of this species is very variable, the females being much larger, the long pubescent antennae of the male and the costate elytra in both sexes will help to distinguish the species which has somewhat the appearance of a *Luperus*.

155. **Dorydea** (*Platyxantha*) **nigripennis**, Jac. — Elongate, subdepressed, fulvous, the breast, abdomen and the apex of the tibiae and the tarsi blackish, thorax bifoveolate, impunctate, elytra black, nearly impunctate.

Mas. Antennae robust, the ninth joint greatly dilated and enlarged, the tenth short, subtriangular, the apex conical, terminal joint longer, dilated and flattened, elytra nearly impunctate, abdominal segments margined with flavous.

Fem. Antennae simple, elytra closely and finely punctured anteriorly, the apex nearly impunctate.

Length 3 $\frac{1}{2}$ lines.

Hab. Si-Rambé.

Of this species, described by myself in the "Notes from the Leyden Mus., 1884" two apparently female specimens are contained in this collection, they agree with the type in most particulars, except that the antennae are entirely fulvous and the tibiae black at the apex as well as the tarsi; another male specimen I have lately received, of which I give the description above which I refer to the present species, on account of its similarity in every respect with the females except in the structure of the antennae, these in their dilated apical joints agree very nearly with the type of the genus *D. insignis*, Baly. In the present insect, the elytra in the male show just a trace of the punctures, when seen under a strong lens, which are so plainly marked in the females, but of course, the possibility is not excluded that the male, here described, may not represent the other sex of the original type, although it is highly probable, and if so, *Dorydea* is a better place for the species than *Platyxantha;* the black elytra will at once separate the insect from *D. insignis*.

156. **Platyxantha submetallica**, n. sp. — Elongate, flavous, the antennae, tibiae and tarsi more or less piceous, thorax bifoveolate, impunctate, elytra finely longitudinally costate, obscure fulvous, the sides and base metallic dark greenish.

Length 3 lines.

Head impunctate, obscure flavous or testaceous, frontal tubercles narrowly transverse, clypeus narrow, triangular, antennae extending to the apex of the elytra, obscure fulvous or fuscous, basal joint elongate, second one very small, the following joints as long as the first one, thorax subquadrate, one half broader than long, the sides straight at the base, rather suddenly rounded before the middle, anterior angles tuberculiform, the disc impunctate, with two large, nearly contiguous foveae, obscure flavous, scutellum triangular, piceous, elytra slightly depressed below the base, each with about eight or nine longitudinal, not highly raised costae, the interstices finely impressed with transversely shaped punctures, dark metallic green, the disc pale obscure flavous, underside and legs flavous, tibiae and tarsi more or less fuscous, the tibiae unarmed, the first joint of the posterior tarsi longer than the following joints together, anterior coxal cavities closed; the penis short and rather broad, its apex broadly rounded, but slightly pointed.

Hab. Si-Rambé.

Whether this species is really distinct from *P. multicostata* is somewhat doubtful, since structural differences seem to be absent, it is however much larger in size, the elytra have the disc pale flavous with but a slight metallic tint and their punctuation is finer, in one specimen the metallic green is replaced by dark purplish. Of both species, two specimens were obtained.

157. **Platyxantha multicostata**, n. sp. — Elongate, testaceous, thorax subquadrate, bifoveolate, impunctate, elytra metallic blue, closely longitudinally costate, the interstices strongly punctured.

Var. The antennae, tibiae and tarsi obscure fuscous.

Length 2 ¾ lines.

Head rather broad, impunctate, the frontal elevations narrowly transverse, clypeus in shape of a highly raised triangular ridge, the central portion extending upwards between the antennae, its anterior margin perfectly straight, labrum broad, flavous, palpi robust, of the same colour, antennae long and slender, extending nearly to the end of the elytra, testaceous, the first joint long, thickened anteriorly, the second one very short, third and following joints extremely long, longer than the basal one, thorax subquadrate, one half broader than long, the sides very slightly rounded before the middle, the angles distinct but not produced, the surface impunctate, minutely granulate, with two deep fovea, separated only by a narrow ridge, scutellum flavous, elytra parallel, with rather feebly raised but closely approached longitudinal ridges, metallic blue or greenish, the interstices closely and strongly impressed with transverse punctures, minutely granulate, underside and legs flavous, the latter slender, unarmed, the first joint of the posterior tarsi longer than the following joints together, claws appendiculate, anterior coxal cavities closed.

Hab. Si-Rambé.

There is no other Malayan species of this genus, which resembles the present one in the costate elytra in connection with its system of coloration.

158. Aenidea costalipennis, n. sp. — Pale flavous or testaceous, head and thorax impunctate, the latter deeply bifoveolate, elytra longitudinally costate, the interstices finely punctured.

Mas. Head deeply excavated below the antennae, palpi incrassate.

Length 3 lines.

Head longer than broad, impunctate, the eyes very large, lower portion of the face deeply excavated, the upper edge of the excavation with two projections, palpi strongly incrassate at the penultimate joint, antennae long and slender, nearly extending to the apex of the elytra, flavous, the first joint long and slender, the second extremely short, the following joints very elongate, terminal joint half the length, thorax scarcely twice as broad as long, the sides nearly straight, slightly widened and rounded before the middle, the disc with two deep fovea, interrupted at the middle, impunctate, elytra minutely granulate, longitudinally costate, the interstices finely transversely punctured, legs long and slender, tibiae with a slender spine, the first joint of the posterior tarsi very long, much longer than the following joints together, claws appendiculate, anterior coxal cavities closed.

Hab. Pangherang-Pisang.

From all other species contained in this genus, the present one is at once distinguished by the costate elytra; the female does not differ except in the structure of the head which is simple.

159. **Syoplia pygidialis**, n. sp. — Elongate, parallel, flavous, the head and the pygidium black, thorax minutely granulate and punctured, elytra pubescent, deeply punctate-striate, the interstices closely longitudinally costate.

Length 2 1/2 lines.

Head broader than long, black, the vertex impunctate, frontal tubercles narrowly trigonate, clypeus broad and flat anteriorly, extending upwards into an elongate point between the antennae, eyes very large, labrum and palpi flavous, the latter incrassate, antennae closely approached at the base, flavous, the first joint extremely long, the second very short, following joints elongate and slender, terminal joints broken off, thorax transverse, of equal width, the sides slightly widened and rounded at the middle, the surface with an obsolete depression near the anterior margin, minutely granulate and very finely and irregularly punctured, flavous, scutellum broadly trigonate, elytra rather convex and parallel, deeply and strongly punctate-striate, the interstices strongly longitudinally costate and clothed with flavous rather long pubescence, apex of the elytra rather truncate, their epipleurae extremely narrow, pygidium black, legs slender, the tibiae with a spine, the first joint of the posterior tarsi, nearly half the length of the tibiae, claws appendiculate, anterior coxal cavities closed.

Hab. Si-Rambé.

From the only other known species of *Syoplia* the present one may be at once known by the black head and pygidium; the pubescent and costate elytra in connection with the long metatarsus of the posterior legs are the principal characters of distinction in this genus.

160. **Doryscus testaceus,** Jac. — Of this extraordinary species, so different in the structure of the claws than the rest of the *Phytophaga*, several specimens were obtained at Si-Rambé; they agree in everything with the type specimen from Ceylon but vary in the colour of the elytra, which latter have in some specimens a narrow transverse piceous band at the base and a small spot near the apex, in others the head and the thorax are more or less marked with piceous and the elytra are narrowly margined with the same colour. To my original description of the genus (Proc. Zool. Soc. 1887) I have to make a correction; in that diagnosis the anterior claws were described as appendiculate and the posterior ones as united at the base, but divided at the apex; in the Sumatran specimen the posterior extremely long and curved claws are widely separated and simple, but with a small tooth at the extreme base, and I can only assume, that in the Ceylon specimen the posterior claws were but accidentally joined; *Doryscus* is so distinct from any other genus of *Galerucinae* that it cannot be mistaken; I have unfortunately not the Ceylon type before me at this moment, but I have no doubt about the identity of the Sumatran insect; the species is also contained in the British Museum from India; the general appearance is that of a *Diabrotica*, the upper and under side are furnished with long bristly hairs, the posterior claws are exceptionally long and curved and their basal portion is as long as the metatarsus, the thorax is deeply impressed at the disc and the elytra are longitudinally costate and pubescent.

161. **Metellus costatipennis,** n. sp. — Black, the head, antennae and thorax fulvous, thorax impunctate, bifoveolate, elytra closely and strongly punctured, the interstices longitudinally costate, black, marked with fulvous below the base, femora fulvous above.

♂. Antennae with the third joint subquadrately and strongly widened, its upper edge deeply emarginate.

♀. Antennae simple, anterior half of the elytra fulvous, posterior one black.

Length 3 ½ lines.

Oblong, rather broad, the head produced, triangularly depres-

sed at the vertex, the latter impunctate, clypeus swollen, very
broad, occupying the entire lower portion of the face, finely
rugose, penultimate joint of the palpi rather swollen, antennae
two-thirds the length of the body, fulvous, the first joint short
and thick, the second one very small, the third greatly enlarged,
elongate and depressed, deeply emarginate at its upper edge,
the other joints elongate, nearly equal, thorax one half broader
than long, the sides constricted at the base, rather suddenly
widened near the middle, the disc rather flattened, with two
small foveae; impunctate, fulvous, scutellum broad, elytra depressed below the base very closely and rather strongly punctured, the interstices longitudinally costate throughout, underside and the legs (the upper edge of the femora excepted) black,
the first joint of the posterior tarsi as long as the following
three joints together, claws appendiculate, anterior coxal cavities
closed.

Hab. Si-Rambé.

Several species belonging to this genus, first described by me
as *Neocharis* and by Baly as *Nacraca*, are now known, the males
are much distinguished by the structure of the antennae (used
as a generic character of doubtful value and more for conve-
nience sake), the present insect differs from any of its allies in
the costate elytra in both sexes which will at once distinguish
it; the system of coloration in the specimens before me varies
greatly, in some, the elytra are black with the anterior portion
of the suture and an indistinct band within the depression, fulvous, in others the latter colour predominates, leaving the posterior half of the elytra black, or the black is reduced to a small
spot below the middle of each elytron.

162. **Sermyloides scutellatus**, n. sp. — Black, the antennae
and legs flavous, head and thorax fulvous, elytra flavous, finely
and closely punctured, the suture and the lateral margin at the
middle, black, last abdominal segment, flavous.

Length 3 $1/2$ lines.

Of convex, elongate shape, the head impunctate, the frontal
elevations very narrowly transverse, lower part paler than the

vertex, palpi robust, antennae rather short (♀) flavous, the first joint slender, curved, the second very small, the third as long as the first joint, terminal joints shorter, thorax transverse, three times broader than long, the anterior margin concave, posterior one parallel to the other, the sides nearly straight, the anterior angles oblique, the surface impunctate, fulvous, scutellum elongate, black, elytra flavous, with a slight purplish gloss, closely and finely punctured, their sides and the epipleurae at the middle black, as well as the extreme sutural margin, the metatarsus of the posterior legs elongate.

Hab. Sumatra (my collection).

This species, of which only a single, apparently female specimen is contained in my collection, differs entirely in its system of coloration, notably the black scutellum, from the few other species of the genus.

163. **Solenia robusta**, n. sp. — Black, the head, thorax and the anterior femora fulvous, thorax extremely minutely punctured, the anterior angles dentiform, elytra metallic dark blue, finely punctate-striate.

Length 3 lines.

Of ovate, convex shape, the head impunctate, deeply grooved between the eyes, the frontal tubercles strongly raised, transverse, palpi incrassate, piceous, antennae scarcely extending to the middle of the elytra, black, the second joint very short, the third, one half shorter than the fourth joint, terminal joints elongate and slender, thorax twice as broad as long, rather convex, the sides with a narrow margin, rounded before the middle, the anterior angles produced into a tooth, the posterior margin with a very short perpendicular notch at each side, the surface scarcely perceptibly punctured, the punctures only visible under a strong lens, scutellum fulvous, elytra convex, very finely punctured in closely approached, somewhat irregular rows, the extreme apex nearly impunctate, underside and legs black, finely clothed with yellow pubescence, the anterior femora fulvous.

Hab. Padang, also Perak.

In the specimens from Perak, contained in my collection, the

four anterior femora are fulvous as well as the first joint of
the antennae; the comparative large and robust size, the fine
punctuation of the elytra and the black antennae and tibiae,
distinguish this species from its allies.

164. Cleonica quadriplagiata, Jac. —— Of this species, three specimens were obtained at Si-Rambé; they differ from the type
in the greater extent of the flavous colour which only leaves a
transverse black band at the base and another below the middle;
in the type the black coloration prevails reducing the flavous
colour to a patch at the middle and another at the apex; the
genus was placed by me in the *Halticinae*, I am however rather
doubtful as to the true position of this insect, as the hind femora are scarcely of sufficient thickness, and the shape and
structure of the head and thorax approaches very near the genus
Xenoda, Baly amongst the *Galerucinae*, the distinct prosternum
however and the robust legs are not peculiar as a rule is this
tribe; be this as it may, *Cleonica* seems to be kind of transition
form between the two great tribes which does not stand alone
in this respect. In the original description of this genus (Notes
Leyd. Mus. Vol. IX) I have omitted to state, that the prosternum
is narrow but distinct.

www.ingramcontent.com/pod-product-compliance
Lightning Source LLC
Chambersburg PA
CBHW021820190326
41518CB00007B/681